Lecture Notes in Networks and Systems

Volume 533

The series "Lecture Notes in Networks and Systems" publishes the latest developments in Networks and Systems—quickly, informally and with high quality. Original research reported in proceedings and post-proceedings represents the core of LNNS.

Volumes published in LNNS embrace all aspects and subfields of, as well as new challenges in, Networks and Systems.

The series contains proceedings and edited volumes in systems and networks, spanning the areas of Cyber-Physical Systems, Autonomous Systems, Sensor Networks, Control Systems, Energy Systems, Automotive Systems, Biological Systems, Vehicular Networking and Connected Vehicles, Aerospace Systems, Automation, Manufacturing, Smart Grids, Nonlinear Systems, Power Systems, Robotics, Social Systems, Economic Systems and other. Of particular value to both the contributors and the readership are the short publication timeframe and the world-wide distribution and exposure which enable both a wide and rapid dissemination of research output.

The series covers the theory, applications, and perspectives on the state of the art and future developments relevant to systems and networks, decision making, control, complex processes and related areas, as embedded in the fields of interdisciplinary and applied sciences, engineering, computer science, physics, economics, social, and life sciences, as well as the paradigms and methodologies behind them.

Indexed by SCOPUS, INSPEC, WTI Frankfurt eG, zbMATH, SCImago.

All books published in the series are submitted for consideration in Web of Science.

For proposals from Asia please contact Aninda Bose (aninda.bose@springer.com).

More information about this series at https://link.springer.com/bookseries/15179

Jan Faigl · Madalina Olteanu ·
Jan Drchal
Editors

Advances in Self-Organizing Maps, Learning Vector Quantization, Clustering and Data Visualization

Dedicated to the Memory of Teuvo Kohonen/Proceedings of the 14th International Workshop, WSOM+ 2022, Prague, Czechia, July 6–7, 2022

Springer

Editors
Jan Faigl
Department of Computer Science
Czech Technical University in Prague
Prague, Czech Republic

Madalina Olteanu
CEREMADE
Université Paris Dauphine-PSL
Paris, France

Jan Drchal
Department of Computer Science
Czech Technical University in Prague
Prague, Czech Republic

ISSN 2367-3370 ISSN 2367-3389 (electronic)
Lecture Notes in Networks and Systems
ISBN 978-3-031-15443-0 ISBN 978-3-031-15444-7 (eBook)
https://doi.org/10.1007/978-3-031-15444-7

This Springer imprint is published by the registered company Springer Nature Switzerland AG
The registered company address is: Gewerbestrasse 11, 6330 Cham, Switzerland

Dedicated to memory of Prof. Teuvo Kalevi Kohonen

Preface

This book contains the peer-reviewed and accepted contributions presented at the 14th International Workshop on Self-Organizing Maps (WSOM+2022) held at the Czech Technical University in Prague (CTU), Prague (Czechia), on July 7–8, 2022. WSOM+2022 is the latest edition of the originally biennial international conferences series. Due to the global pandemic experienced in the last years, the 14th edition has been postponed to follow the tradition of in-person discussions among the participating researchers. A quarter of a century after the first conference WSOM'97 organized in Helsinki, Finland, with Prof. Teuvo Kohonen as General Chairman, we meet in Prague. Since that, self-organizing maps and related unsupervised learning methods have been developed, showing theoretical results and wide adaption in many practical deployments. In 2022, it is very sad that Prof. Kohonen passed in the fall of 2021. Therefore, we would like to express our gratitude for his active and fruitful research career and for founding the unsupervised learning of self-organizing maps. It was originally intended as a technique for projecting high-dimensional problems into a two-dimensional space, supporting visualization, and understanding the problem. Over more than twenty-five years, self-organizing maps have been utilized in problems that were probably not thought of at the beginning. Professor Kohonen, thank you for all your contributions that enrich our society and research field; WSOM+ 2022 is dedicated to you!

In this collection, the reader can find all sorts of advancements in self-organizing maps (SOMs) and learning vector quantization (LVQ), including progressive ideas on exploiting features of parallel computing. The collection is balanced in presenting novel theoretical contributions with applied results in traditional fields of SOMs, such as visualization problems and data analysis. Furthermore, the collection also includes less traditional deployments in trajectory clustering and advancements in exploiting quantum computation. The presented book is worth interest to data analysis and machine learning researchers and practitioners, specifically those interested in being updated with current developments in unsupervised learning, data visualization, and self-organization.

The book collects the results of about thirty researchers from eight countries. It is thus a result of all the collective effort that would not be possible without the advice and guidance of the international WSOM Steering Committee. Furthermore, the final revisions of the accepted papers are also a result of the reviews by the Program Committee members and additional anonymous reviewers.

May 2022

Organization

Steering Committee

Marie Cottrell, France
Pablo Estevez, Chile
Jean-Charles Lamirel, France
Thomas Martinetz, Germany
Erzsebet Merenyi, USA
Madalina Olteanu, France
Alfredo Vellido, Universitat Politécnica de Catalunya, Spain
Michel Verleysen, Belgium
Thomas Villmann, Germany
Takeshi Yamakawa, Japan
Hujun Yin, UK

Program Committee

Guilherme Barreto	Universidade Federal do Ceará, Brazil
Cyril de Bodt	Université catholique de Louvain, Belgium
Marie Cottrell	SAMM Université Paris 1 Panthéon-Sorbonne, France
Jan Drchal	Czech Technical University in Prague, Czechia
Jan Faigl	Czech Technical University in Prague, Czechia
Hervé Frezza-Buet	CentraleSupélec, France
Tina Geweniger	Westsaechsische Hochschule Zwickau, Germany
Yann Guermeur	CNRS, France
Pitoyo Hartono	Chukyo University, Japan
Martin Holena	Czech Academy of Sciences, Czechia
Michael Kirby	Colorado State University, USA
Markus Koskela	CSC - IT Center for Science Ltd., Finland
Jean-Charles Lamirel	Inria Project TALARIS-LORIA, France
Mustapha Lebbah	Université Paris 13, LIPN-CNRS, France
John Lee	Université catholique de Louvain, Belgium

Victor Lobo	CINAV-Escola Naval, Portugal
Mike Mendenhall	Air Force Research Laboratory, USA
Madalina Olteanu	Paris Dauphine PSL University, France
Andreas Rauber	Vienna University of Technology, Austria
Fabrice Rossi	Université Paris Dauphine, France
Nicolas Rougier	Inria, France
Frank-Michael Schleif	University of Applied Sciences Wuerzburg-Schweinfurt, Germany
Udo Seiffert	Compolytics, Germany
Alfredo Vellido	Universitat Politécnica de Catalunya, Spain
Michel Verleysen	Universite catholique de Louvain, Belgium
Thomas Villmann	University of Applied Sciences, Mittweida, Germany
Axel Wismueller	University of Rochester, USA
Hujun Yin	The University of Manchester, UK

Contents

Sparse Weighted K-Means for Groups of Mixed-Type Variables

Marie Chavent[1], Marie Cottrell[3], Jerome Lacaille[2], Alex Mourer[1,2,3], and Madalina Olteanu[4(✉)]

[1] Inria Bordeaux Sud-Ouest, Équipe ASTRAL-IMB, UMR CNRS 5251, U. Bordeaux, Talence, France
[2] Safran Aircraft Engines-Datalab-Villaroche, Moissy Cramayel, France
[3] SAMM, EA 4543 - Université Paris 1 Panthéon-Sorbonne, Paris, France
[4] CEREMADE, UMR 7534 - Université Paris Dauphine PSL, Paris, France
olteanu@ceremade.dauphine.fr

Abstract. Assessing the underlying structure of a dataset is often done by training a clustering procedure on the features describing the data. In practice, while the data may be described by a large number of features, only a minority of them may be actually informative with regard to the structure. Furthermore, redundant features may also bias the clustering, whether one speaks of redundancy in the informative or the uninformative features. The present contribution aims at illustrating two sparse clustering methods designed for mixed data (made of numerical and categorical features). The proposed methods summarise redundant features into groups, and select the most relevant groups of features only in the clustering procedure. The performances and the interpretability of the sparse methods are illustrated on a real-life data set.

Keywords: Sparse clustering · Feature clustering · Feature selection · Group of features selection · Variable importance

1 Introduction

Clustering is one of the most common tasks in unsupervised learning. In the case of vector data, it consists in partitioning observations into homogeneous groups, described by the available features. When the number of features is large, and this is more and more the case with contemporary massive data collection, it is legitimate to assume that clusters differ according to a limited number of features only. These *important* features are the only actually informative about the underlying structure of the data. The others, called *noise* features, are not only uninformative, but they may actually dilute or even *mask* the structure in the data. For this reason, several sparse clustering methods have been developed in the literature [5,6,8], and aim at performing clustering by selecting and using the important variables only.

Nevertheless, distinguishing informative from uninformative features is not the only issue in high dimensional clustering. Feature redundancy may also have an impact on clustering performances. Indeed, if several features bring similar

J. Faigl et al. (Eds.): WSOM+ 2022, LNNS 533, pp. 1–10, 2022.
https://doi.org/10.1007/978-3-031-15444-7_1

information about the data, then one may wish either to select them together as informative, or drop them together as noise. Sparse clustering should thus be combined with introducing a group structure on the features, and performing group of features selection for clustering. If the groups of features are priorly known - this may be the case for repeated measures, categories of features, etc. - then one may apply a group-sparse clustering procedure, such as the one proposed in [3]. If the groups are not priorly known, which is most often the case, they may be built using a feature clustering algorithm which groups them according to their correlations, as in [2].

One should remark here that the above is a different paradigm from that of co-clustering [4]. Co-clustering focuses on simultaneously clustering features and observations, and on partitioning the tabular data into blocks both on the rows and the columns. In our case, we wish to cluster the observations using the most informative features only, and taking into account the possible redundancy between them.

The present manuscript focuses on sparse-group clustering, where the groups of features are built using the correlations between them. One starts by performing a hierarchical clustering on the features, using ClustOfVar algorithm [2]. Second, one trains a sparse clustering on the data, using the feature group structure. Two different approaches will be investigated here: either use groups of features directly and apply a group-based penalty as in [3], or summarise each group of features by a synthetic variable and apply the usual sparse clustering as in [8].

Eventually, let us mention here that since tabular data often contains mixed information (numerical and categorical features), the proposed methodology should be able to handle it properly. The usual approach would be to use one-hot encoding and then process the resulting dummies as numerical features. However, this processing step is generally innapropriate since the resulting Euclidean distance is not compatible without proper scaling. Furthermore, the dummy variables resulting from a categorical feature should be treated as a block and not be split during the feature clustering step. For these two reasons, in the case of mixed data, a preprocessing step based on [1] will be systematically applied.

The rest of the paper is organised as follows. In Sect. 2, we recall the procedures for preprocessing mixed data and for building groups of features based on their correlations, as well as the sparse k-means procedure with ℓ_1 and ℓ_1-group penalties. Section 3 summarises the two different approaches for sparse clustering in case of redundant features, by either preserving whole groups of features, or summarising them using synthetic variables. Section 4 contains a detailed illustration of the proposed methodology on a real-life data set.

2 Methods

2.1 Redundancy-Based Feature Clustering

We place ourselves in the case where the data contains possibly correlated and redundant features, which may have a negative impact on observation clustering.

In order to identify them, groups of related and redundant features may be computed using, for instance, the ClustOfVar algorithm [2], that we briefly recall below.

Suppose one has a tabular dataset, made of n individuals and p features, $\mathbf{X} \in \mathbb{R}^{n \times p}$. Furthermore, suppose the data is mixed, such that the first p_1 features are categorical, with d_j categories each, for $j = 1, \ldots, p_1$, and the last p_2 features are numeric, $p = p_1 + p_2$. One may alternatively use the notation $\mathbf{X} = [\mathbf{X}_1 | \mathbf{X}_2] \, \mathbb{R}^{n \times (p_1 + p_2)}$.

ClustOfVar performs a hierarchical clustering analysis (HCA) of the available features. At each aggregation step, two groups of features $\mathbf{G}_\ell$ and $\mathbf{G}_{\ell'}$ are joint together if they minimise the following dissimilarity criterion:

$$d(\mathbf{G}_\ell, \mathbf{G}_{\ell'}) = H(\mathbf{G}_\ell) + H(\mathbf{G}_{\ell'}) - H(\mathbf{G}_\ell \cup \mathbf{G}_{\ell'}), \tag{1}$$

where

$$H(\mathbf{G}_\ell) = \sum_{X_j \in \mathbf{X}_1 \cap \mathbf{G}_\ell} \frac{\sum_{s=1}^{d_j} n_{j,s} (\overline{F}_{\ell,s} - \overline{F}_\ell)^2}{\sum_{i=1}^{n} (F_{\ell,i} - \overline{F}_\ell)^2} + \sum_{X_j \in \mathbf{X}_2 \cap \mathbf{G}_\ell} \operatorname{cor}(F_\ell, X_j)^2 \tag{2}$$

quantifies the homogeneity of the group of features $\mathbf{G}_\ell$. In the above equation, F_ℓ is the first principal component stemming from a mixed principal component analysis (mixed PCA, [1]) restricted to the features in group $\mathbf{G}_\ell$, $F_{\ell,i}$ is the ith observation of the feature F_ℓ and $n_{j,s}$ is the number of input data taking the sth value of the jth categorical feature. For clarity, $\overline{F}_\ell$ (or $\overline{F}_{\ell,s}$) denotes the average of the feature F_ℓ (or the average of the feature F_ℓ restricted to the observations belonging to the sth category of the feature X_j).

One may remark that the terms in the left-hand sum above are correlation ratios between the categorical features X_j and the numerical synthetic feature F_ℓ, and that the terms in the right-hand sum are squared Pearson correlations between the numerical features X_j and F_ℓ. By construction, F_ℓ represents the synthetic feature most linked to all features in group $\mathbf{G}_\ell$, hence it represents a good summary of the group.

Cutting the resulting dendrogram produces a partitioning of the features into L groups $\mathbf{G}_1, \ldots, \mathbf{G}_L$, each group being homogeneous in terms of redundancy of information. At this point, one may either use an index labelling each feature X_j according to its group, or summarise each group $\mathbf{G}_\ell$ by the first principal component of the mixed PCA trained with the features in the group, F_ℓ. The latter would result into a reduced data set $\mathbf{F} = [F_1, \ldots, F_L]$.

The issue of selecting L, the number of groups of features, is not discussed here. In practice, usual criteria for selecting the number of clusters may be used.

2.2 Mixed Data Pre-processing

The first step of the preprocessing consists in transforming each categorical feature X_j into d_j dummy variables $\tilde{\mathbf{X}}_j = (\tilde{X}_j^1, \ldots, \tilde{X}_j^{d_j}) \in$

$\{0,1\}^{n \times d_j}$ and thus define a block structure on the transformed data $\tilde{\mathbf{X}} = \left[\tilde{\mathbf{X}}_1|\ldots,|\tilde{\mathbf{X}}_{p_1}|X_{p_1+1}|\ldots,|X_{p_1+p_2}\right]$.

The second step consists in scaling the data $\tilde{\mathbf{X}}$ into a matrix $\mathbf{Z}$, as described for instance in [1]: numerical variables are scaled to zero mean and unit variance, while dummy variables are centred and scaled by $\sqrt{\frac{n_{j,s}}{n}}$. The scaling applied to the dummy variables actually leads to using a χ^2 distance on the categorical features, while the numerical features, after scaling, are compared through the usual Euclidean distance. Eventually, let us also mention that this step is somehow independent from the feature clustering above. Nevertheless, the preprocessing will appear useful in the clustering step that we explain next.

2.3 Sparse Weighted *k*-Means

k-means partitioning is one of the most popular clustering algorithms, its popularity being supported especially by its simplicity: for a given positive integer K, one searches the *best* partitioning of the data $\mathbf{Z}$ into K clusters $C_1,\ldots,C_K$ where $C_1,\ldots,C_K$ denotes a partitioning of the observation indices $\{1,\ldots,n\}$. *Best* stands here for homogeneous clusters in terms of within-class variance. In practice, this amounts to finding $C_1,\ldots,C_K$ maximising the between-class variance:

$$\underset{C_1,\ldots,C_K}{\text{maximise}} \sum_{j=1}^{d} b_j = \underset{C_1,\ldots,C_K}{\text{maximise}} \sum_{j=1}^{d}\sum_{k=1}^{K} \frac{n_k}{n}(\overline{Z}_{j,k} - \overline{Z}_j)^2, \tag{3}$$

where $b_j = b_j(\mathbf{Z}, C_1,\ldots,C_K)$ is the between-class variance on the jth feature of the recoded data $\mathbf{Z}$, $d = d_1 + \ldots + d_{p_1} + p_2$, and $\overline{Z}_j$ (resp. $\overline{Z}_{j,k}$) is the average of feature Z_j (resp. the average of feature Z_j restricted to the observations belonging to the kth cluster).

Since *k*-means does not allow for feature selection, several extensions have been developed in the literature as a solution [3,8]. These extensions build on the idea of regularised models, and introduce a vector of weights associated to the features, and a penalty function applied to the weighted between-class variance. More formally, one aims at solving a new optimisation problem, by seeking the vector of non-negative weights $\mathbf{w} = (w_1,\ldots,w_d)^\intercal$ and the partitioning $C_1,\ldots,C_K$ maximising the penalised criterion:

$$\underset{\mathbf{w},C_1,\ldots,C_K}{\text{maximise}} \sum_{j=1}^{d} w_j b_j - \lambda h(\mathbf{w}), \text{ subject to } \|\mathbf{w}\|_2^2 \leq 1,\ w_j \geqslant 0\ \forall j, \tag{4}$$

where $h(\mathbf{w})$ is a non-negative penalty function assessing the complexity of the model, and $\lambda \geqslant 0$ is a hyper-parameter controlling the strength of the penalty.

Since the optimisation problem above is discrete and non-convex, an iterative algorithm may be used for the numerical approximation of the solution. For a fixed number of clusters K, a fixed tuning parameter λ, and a penalty term $h(\mathbf{w})$, the optimisation problem above may be solved using an iterative algorithm, which alternates two steps:

1. for a fixed $\mathbf{w}$, find $C_1, \ldots, C_K$ maximising $\sum_{j=1}^{d} w_j b_j$. This amounts to training usual k-means on the scaled featurres $\tilde{Z}_j = \sqrt{w_j} Z_j$, $j = 1, \ldots, d$.
2. for a fixed partitioning $C_1, \ldots, C_K$, find $\mathbf{w}$ maximising the criterion in Eq. 4. Depending on the penalty term, analytical solutions may be fully written, as for instance in [3,8].

For more details on the initialisation of the algorithm, the stopping criterion, or the choice of the hyper-parameter term λ, the reader may refer again to [8] or [3]. Two penalty functions will be particularly of interest in the following, the ℓ_1 norm, and the ℓ_1-group norm.

Weighted Sparse k-Means with ℓ_1-Penalty. If all the features in the data set are numerical ($d = p$ and $\mathbf{Z}$ is the scaled matrix $\mathbf{X}$), and if one wishes to perform feature selection without supposing any group structure on the latter, the penalty function may be chosen as:

$$h(\mathbf{w}) = \|\mathbf{w}\|_1 = \sum_{j=1}^{p} w_j. \tag{5}$$

This penalty was introduced in [8], by analogy with the *lasso* regression framework [7]. This penalty will drop all features which between-class variance is smaller than λ.

Weighted Sparse k-Means with ℓ_1-Group Penalty. If the features in the data set $\mathbf{Z}$ are available as groups $\mathbf{G}_1, \ldots, \mathbf{G}_L$ and labelled accordingly as explained in Sect. 2.1, the penalty term should be chosen so that it either keeps, or discards a whole group of features. The ℓ_1-group penalty, first used in a regression framework by [9], was introduced in the sparse k-means framework by [3]. The penalty writes as:

$$h(\mathbf{w}) = \sum_{l=1}^{L} \sqrt{|\mathbf{G}_\ell|}\|\mathbf{w}_\ell\|_2, \tag{6}$$

where $|\mathbf{G}_\ell|$ is the number of features in $\mathbf{Z}$ belonging to group $\mathbf{G}_\ell$, and $\mathbf{w}_\ell$ is a vector containing the weights associated to the features in group $\mathbf{G}_\ell$. One should remark here that each categorical feature X_j has been processed as d_j scaled dummies in $\mathbf{Z}$, hence it will be associated to one group $\mathbf{G}_\ell$ only, and will contribute with d_j features to the size of the group.

In particular, since the pre-processing of the mixed data in Sect. 2.2 generates a block structure for the processed features, each categorical feature X_j in $\mathbf{X}$ being a block of size d_j in $\mathbf{Z}$, the ℓ_1-group penalty may be naturally used for feature selection in clustering with mixed data. This framework has been investigated in [3].

3 Two Methods for Sparse Clustering in the Presence of Feature Redundancy

Using the tools introduced in the previous section, one may now build two methods for sparse k-means clustering for mixed data, and in the presence of correlated or redundant features. In both cases, one will start by computing groups of redundant features using ClustOfVar. Next, she will either use the resulting group structure with the ℓ_1-group penalty in the weighted sparse k-means procedure, or she will focus on the synthetic features summarising the groups of features, and train weighted sparse k-means with the ℓ_1-penalty.

Algorithm 1 below summarises the first approach. As one may see, this approach implies using the entire data set $\mathbf{Z}$ in the clustering procedure. For the selected groups of features, the weights of all the features in those groups will be available for further processing and interpretation.

The second method is summarised by Algorithm 2. In this framework, each group of features $\mathbf{G}_\ell$ is summarised by a synthetic variable F_ℓ. This implies on the one hand some loss of information since clustering is not trained on the full data set, and on the other hand a simpler and faster approach. Indeed, the dimensionality reduction obtained through the synthetic variables translates into lower sparse k-means complexity, and better interpretability of the weights $\mathbf{w}$.

Algorithm 1. Sparse-group k-means clustering

1: Train ClustOfVar on the mixed data set $\mathbf{X}$.
2: Use the resulting dendrogram and select the number of groups of features L.
3: Extract the group labels for all features.
4: Compute the transformed and scaled data $\mathbf{Z}$.
5: Associate a group label to each transformed feature in $\mathbf{Z}$ (in particular, scaled dummies stemming from the same categorical feature will have the same label).
6: Perform weighted sparse k-means with the ℓ_1-group penalty on the groups of transformed features in $\mathbf{Z}$.
7: Tune K and λ in order to have meaningful clusters, and a good trade-off between the ratio of explained variance and the parsimony in terms of selected groups.

Algorithm 2. Sparse k-means clustering on synthetic features

1: Train ClustOfVar on the mixed data set $\mathbf{X}$.
2: Use the resulting dendrogram and select the number of groups of features L.
3: Extract the synthetic features $F_1, \ldots, F_L$.
4: Perform weighted sparse k-means with the ℓ_1-penalty on the synthetic features matrix $\mathbf{F}$.
5: Tune K and λ in order to have meaningful clusters, and a good trade-off between the ratio of explained variance and the parsimony in terms of selected groups.

4 Experimental Results

The two algorithms will be illustrated on a wine dataset[1]. The dataset contains 21,605 observations, each of them representing a bottle of wine coming from different regions of the world. We decided to focus on French wines only. The data selected for illustration contains 3,368 bottles described by 29 mixed-type

[1] https://www.kaggle.com/datasets/dev7halo/wine-information?select=cleansingWine.csv.

features of which one only is numeric: the price. Five categorical features qualify the wines by sweetness, acidity, body, level of tannin (each of them has 5 levels), and by their type (red, white, rose, sparkling). The remaining 23 features indicate the presence/absence of the grape variety such as Merlot, Chardonnay, Syrah, etc. In addition, the variable representing the wine regions was not included as an input feature, but will be used as a control variable.

Data is first plotted on the first two principal axes of a PCAmix[2]. In Fig. 1(a), axis 1 opposes red and tannic wines (left) to white and not tannic ones (right). Sweet wines are in the upper part of the graph, while dry wines are in the lower part. The colours in Fig. 1(b) stand for the different wine regions such as Champagne (green) or Bordeaux (orange).

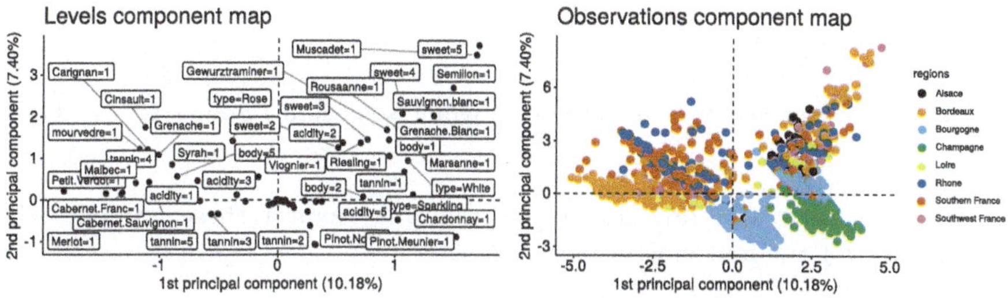

(a) Features (b) Wines coloured by region

Fig. 1. PCAmix on the wine data. First two principal components.

We now look for a partitioning into $K = 4$ clusters of the wine data. We start by extracting groups of redundant features using the ClustOfVar algorithm as explained in Sect. 2.1. Figure 2 represents the ClustOfVar dendrogram. $L = 7$ groups of features are extracted, and are used as input for Algorithms 1 and 2.[3]

Results of Algorithm 1. Figure 3 summarises the output of Algorithm 1. Figure 3(a) gives the regularisation paths of the group weights $\|\mathbf{w}_\ell\|_2$ according to the hyper-parameter λ. For each value of λ, the group-sparse k-means algorithm finds a partitioning into 4 clusters and selects less groups of features as λ increases. Figure 3(b) shows the associated proportion of explained variance, computed on the entire set of features. One will look for a trade-off between a large explained variance and a small set of selected groups. In this case, $\lambda \simeq 0.24$ preserves the same amount of explained variance as the configuration without any sparsity, while reducing the number of selected groups to 3 ($\mathbf{G}_1$, $\mathbf{G}_5$, $\mathbf{G}_7$). The features in $\mathbf{G}_1$ are essentially varieties specific to Rhône and Southern regions, those in $\mathbf{G}_5$ are white grape varieties specific to Champagne and Bourgogne regions as well as type and tannin level, while $\mathbf{G}_7$ is made

[2] Implemented in the R-package `PCAmixdata`.

[3] The sparse clustering methods are implemented in the R-package `vimpclust` (https://cran.r-project.org/web/packages/vimpclust/index.html) and ClustOfVar is implemented in the `ClustOfVar` R-package.

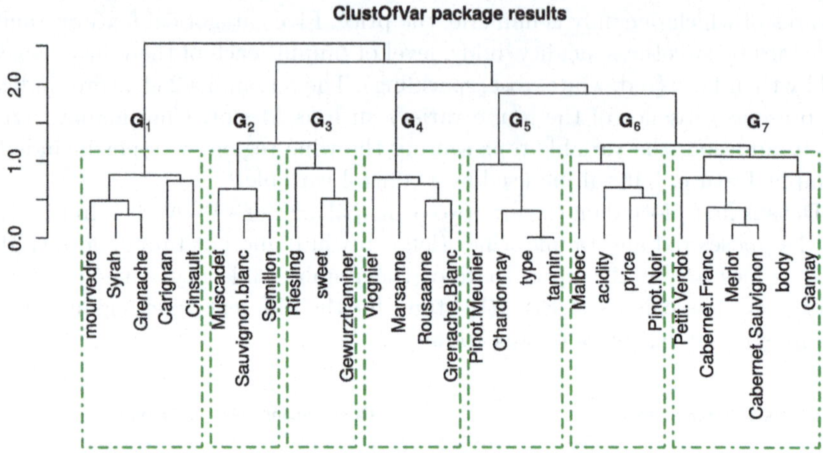

Fig. 2. ClustOfVar dendrogram. The dendrogram is cut into 7 groups $\mathbf{G}_1, \ldots, \mathbf{G}_7$.

of varieties specific to Bordeaux region. One may expect these groups to be discriminant for the four-cluster partitioning.

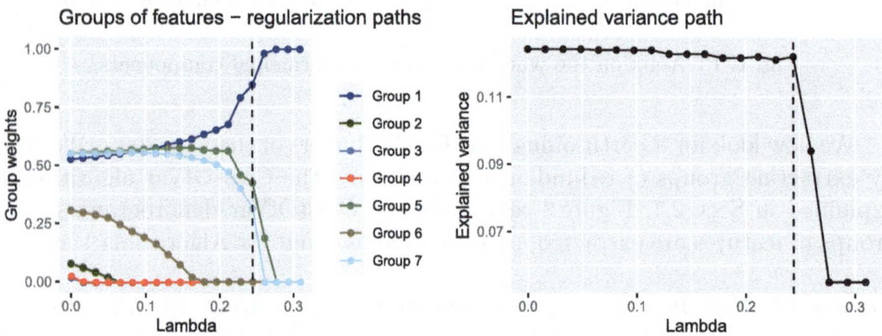

(a) Group weights given λ

(b) Explained variance given λ

Fig. 3. Results of Algorithm 1.

Results of Algorithm 2. The results of Algorithm 2 are summarised in Fig. 4. In this case, regularisation paths in Fig. 4(a) contain the weights w_ℓ associated to the synthetic numerical features F_ℓ. The explained variance in Fig. 4(b) appears larger than in Fig. 3(b), but one should recall that the dimensionality of the original data has been reduced here to $L = 7$ thanks to the use of synthetic variables. One should remark that the selected synthetic variables correspond to exactly the same groups selected with Algorithm 1, although not exactly in the same order. On the other hand, the adjusted Rand index between the clusterings obtained with the two methods is equal to 0.93. We may thus conclude the two methods are globally coherent. Nevertheless, Algorithm 2 is much faster, and clusters can be easily interpreted through the selected synthetic variables and their relationship to the initial features.

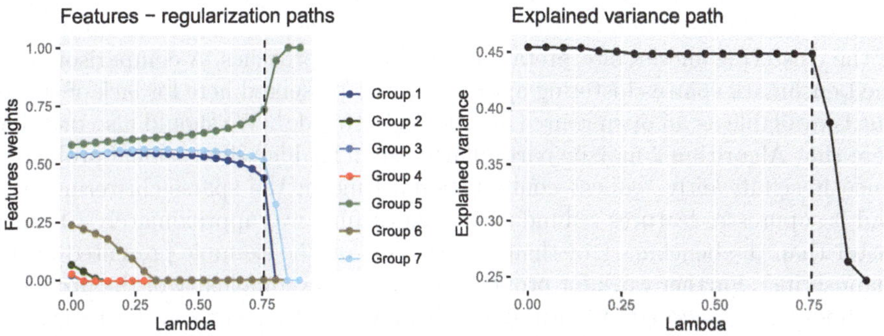

(a) Synthetic variables weights given λ (b) Explained variance given λ

Fig. 4. Results of Algorithm 2.

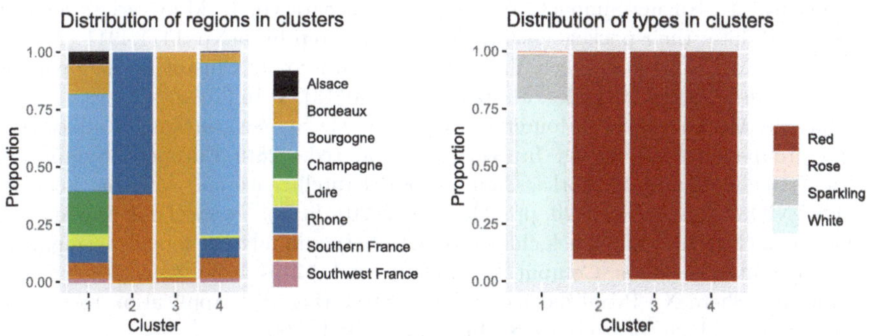

Fig. 5. Description of the clusters computed with Algorithm 2.

In Fig. 5, we illustrate clustering relevance by studying region and type distribution in the partitioning obtained with Algorithm 2. The four clusters obtained with the synthetic features summarising groups $\mathbf{G}_1$, $\mathbf{G}_5$, and $\mathbf{G}_7$ are homogeneous both from type and region point of view. Cluster 1 contains essentially white and sparkling wines from all regions. Cluster 2 contains red and rosé wines from the Rhône and Southern France regions. Cluster 3 is associated to red Bordeaux wines, while cluster 4 is mainly associated with red Bourgogne wines.

5 Conclusion

The present contribution introduced a method combining variable clustering through ClustOfVar and sparse weighted k-means, the latter applied either to groups of features, or to synthetic features summarising the groups of features. Thanks to the illustration on real-life data, one may assess the good performances of the clustering and its interpretability for both approaches. A comparison with the benchmark sparse clustering approach is not presented here for lack of space, but is available as an open source notebook on Kaggle[4]. We should also mention here that Algorithm 2 may be particularly useful for high-dimensional problems, since it significantly reduces computational time in the sparse-clustering step and it suppresses features redundancy (Algorithm 2 was approximately 10 times faster than the benchmark or than Algorithm 1 on the example presented in the manuscript). Further work for properly assessing the complexity of the proposed methodology is needed. We are currently investigating both an evaluation of the method using various synthetic simulated data and the available state-of-the art on sparse clustering, and a thorough study of the complexity, including ClustOfVar step.

References

1. Chavent, M., Kuentz-Simonet, V., Labenne, A., Saracco, J.: Multivariate analysis of mixed data: the r package pcamixdata. arXiv preprint arXiv:1411.4911 (2014)
2. Chavent, M., Kuentz-Simonet, V., Liquet, B., Saracco, J.: Clustofvar: an r package for the clustering of variables. J. Stat. Softw. **50**(13), 1–16 (2012)
3. Chavent, M., Lacaille, J., Mourer, A., Olteanu, M.: Sparse k-means for mixed data via group-sparse clustering. In: Verleysen, M. (ed.) 28th European Symposium on Artificial Neural Networks, Computational Intelligence and Machine Learning (ESANN), 2–4 October 2020, pp. 235–240 (2020). https://www.i6doc.com/en/
4. Govaert, G., Nadif, M.: Block clustering with bernoulli mixture models: comparison of different approaches. Comput. Stat. Data Anal. **52**(6), 3233–3245 (2008)
5. Pan, W., Shen, X.: Penalized model-based clustering with application to variable selection. J. Mach. Learn. Res. **8**(May), 1145–1164 (2007)
6. Raftery, A., Dean, N.: Variable selection for model-based clustering. J. Am. Stat. Assoc. **101**(473), 168–178 (2006)
7. Tibshirani, R.: Regression shrinkage and selection via the lasso. J. Roy. Stat. Soc. Ser. B (Methodol.) **58**(1), 267–288 (1996)
8. Witten, D., Tibshirani, R.: A framework for feature selection in clustering. J. Am. Stat. Assoc. **105**(490), 713–726 (2010)
9. Yuan, M., Lin, Y.: Model selection and estimation in regression with grouped variables. J. Roy. Statist. Soc. Ser. B (Statist. Methodol.) **68**(1), 49–67 (2006)

[4] https://www.kaggle.com/code/alexpoglaghi/wsom2022.

Fast Parallel Search of Best Matching Units in Self-organizing Maps

Yann Bernard(✉) and Bernard Girau

Université de Lorraine, CNRS, LORIA, 54000 Nancy, France
{Yann.Bernard,Bernard.Girau}@loria.fr

Abstract. Self-Organizing Maps (SOM) are unsupervised neural networks that map an underlying regular neighbourhood structure onto a codebook that is learned to perform vector quantization onto an input space. They are used in a wide range of applications, where an increase in the number of neurons in the SOM often leads to better results or new emerging properties. Therefore highly efficient algorithms for learning and evaluation are key to improve the performance of such models. The most time-greedy component of the SOM learning algorithm is the computation of the best matching unit (BMU) that is usually performed by means of a standard Winner Takes All algorithm. In a previous paper, we have proposed FastBMU, an algorithm to compute the BMU that scales better with a large number of neurons. Our algorithm has shown a significant improvement in computing time with a minimal degradation of performance in the context of a sequential implementation. In the perspective of hardware implementations, we propose here a parallel version of FastBMU, and we analyse its behavior and its performance. Based on the performed analysis, we finally derive principles of a parallel hardware structure that maximize resource utilization.

Keywords: Self-Organizing Maps · vector quantization · parallel computation

1 Introduction

A hardware implementation on FPGA of self-organizing maps would require parallelism to efficiently use the available resources. Typical solutions, reviewed in [5], feature either calculating all the distances in parallel, which is inefficient, or using tree-search algorithms [2], which does not fit our bio-inspired cellular approach. Therefore we propose to adapt the FastBMU algorithm to parallel processing. FastBMU is however a deeply sequential algorithm: each step depending heavily on the previous one. Making a parallel version of it would require a fundamental change in its operation. A simpler would be to parallelize at the level of the iterations instead of the level of the BMU search. Indeed, if it is possible to parallelize the iterations in a way that allows the use of almost all the neurons simultaneously with FastBMU without compromising the quality of the BMUs found, then we have a way to be more efficient than the classically parallel versions of SOM that compute all the neuronal distances at the same time, because FastBMU does far fewer evaluations than the standard version.

J. Faigl et al. (Eds.): WSOM+ 2022, LNNS 533, pp. 11–20, 2022.
https://doi.org/10.1007/978-3-031-15444-7_2

2 Self-organizing Maps

2.1 Vector Quantization

Vector quantization (VQ) consists in approximating the probability density of an input space (often known through a large set of samples) with a finite set of prototype vectors, or codewords. The set of codewords is called the codebook. Many VQ techniques exist such as k-means [8], self-organizing maps (SOM) [6], neural gas (NG) [10], growing neural gas (GNG) [4], or growing when required (GWR) [9]. The well-known k-means algorithm only performs VQ, while the other ones simultaneously map a neighbourhood structure onto the codebook, i.e., connections between codewords. Algorithms such as NG, GNG, or GWR present good performance in terms of minimization of the quantization error, measured by the mean squared error. However the performance is often related to the creation in the network of a very large number of new prototypes and/or connections between prototypes, since the neighbourhood structure is learned together with the codewords. This induces a significant increase in computational cost. On the contrary SOM are based on a static underlying topology and a fixed number of codewords associated to neurons that have a predefined position in a map of the predefined topology (usually a 2D lattice). The spatial arrangement of the prototypes in the map makes it able to capture the topographic relationships (i.e., the similarities) between the inputs: similar blocks of samples tend to be represented by spatially close neurons in the map. The SOM model is also a well-known and biologically plausible model of the topographical mapping of the visual sensors onto the cortex [11].

2.2 Kohonen SOM

In a Kohonen SOM [7], each neuron n is connected to the d-dimensional input and has a d-dimensional weight vector w_n, or codeword. The neurons are spatially arranged in a discrete map, usually a 2D grid, or a hexagonal tiling.

All codewords are initialized with random weights. The training lasts several epochs. One epoch includes enough training iterations such that the whole training dataset has been used once for learning. For each iteration, a training vector v is picked among the inputs. All neurons n compute the distance (usually L^2) between w_n and v. The neuron with the minimal distance is called the best matching unit (BMU). The weights are then updated according to:

$$w_i(t+1) = w_i(t) + \epsilon(t) \cdot \Theta(\sigma(t), d_{i,BMU}) \cdot (v - w_i(t)) \tag{1}$$

where ϵ and σ are time functions[1], and $d_{i,BMU}$ is the normalized distance between neuron i and the BMU in the map (not in the input space). Θ is a nor-

[1] $\sigma(t)$ influences the neighbourhood function. The higher it is, the more the BMU influences other neurons. Here, it starts at 0.5 and linearly decreases to 0.001, so that at the beginning of the training, neurons are significantly influenced by the BMU (unfolding the SOM), and at the end, nearly none except the BMU are (optimizing the quantization). $\epsilon(t)$ is the learning parameter, it starts at 0.6 and linearly decreases to 0.05. We ran the SOM for 10 epochs.

malized centered Gaussian function with standard deviation σ. During learning, the SOM gradually unfolds, thus mapping the input space onto the underlying structure of the SOM, while preserving the probabilistic properties of the input space (i.e., with more prototypes in the regions of higher density).

An iterative batch version of SOM learning exists, but it is far less used than the above "on-line" version, and it does not fit applications where the dataset dynamically evolves. Therefore we present our efficient method to compute the BMU in the context of the standard on-line learning algorithm, as in [3].

3 FastBMU

We describe here the sequential FastBMU algorithm that has been used as the basis for our parallel version. More details, like computational speed-up, can be found in [3]. Its goal is to replace the traditional exhaustive BMU search with a more computationally efficient gradient descent. FastBMU takes advantage of the dimensionality reduction property of the SOM: close neurons in the SOM topology represent close vectors in the input space. Therefore when an input vector is presented to the SOM, a pseudo-continuous gradient appears in the SOM when considering the distances between the neuron weights and the input vector whose minimum is located at the BMU.

Algorithm 1. Particle

Input: *pos* : current position
Parameters: *values*: global table of distances, *eval*: global history of evaluations
Output: *bmu* : best matching unit's index

if *eval(pos)* is True **then**
    return *pos*
end if
eval[pos] ← True
Let *new_pos* take the index of the closest not evaluated neighbour to the input vector (ignoring neighbours *p* such that *eval(p)* =True)
if *values[new_pos]* ≤ *values[pos]* **then**
    return Particle(*new_pos*)
end if
return The index of the smallest value between *pos* and the returned indexes by execution of Particle() on all direct neighbours of *pos*.

The core of FastBMU is a discretized gradient descent on the neurons that minimizes the distance between them and the input vector. Algorithm 1 shows the pseudo-code of the gradient descent, called *Particle*. A *Particle* starts at an arbitrarily chosen position in the map. In the example illustrated in Fig. 1, it starts at coordinates (0,0) (upper left corner). The selected neuron has two neighbors, one to the east (1,0) and one to the south (0,1). We evaluate the distances to the input vector of these three neurons ((0,0), (0,1) and (1,0)).

If one of the neighboring neurons gives the smallest distance, then the gradient descent goes towards this neuron, which becomes the next selected position from which we repeat the whole process. If the smallest distance is found where we are currently positioned, then we consider that we are in a local minimum. In our example, the smallest distance is at position (1,0). The search process therefore moves one step eastward. The next neuron has three neighbors, (1,1), (2,0) and (0,1). The latter is ignored since we have just come from it. We compare the three distances ((1,0), (1,1) and (2,0)) again and move to the lowest distance (southward). The process iterates until it reaches a local minimum at position (5,6), where all neighboring neurons have a higher distance to the input vector than the selected neuron.

However, finding a neuron without better neighbors does not guarantee that no other better minimum can be found locally. For example in Fig. 1 the neurons south and west of the first local minimum have a gradient that points to another neuron than the current local minimum. A possible cause, especially for grid topologies, is that the gradient can be locally diagonally oriented, and therefore here it would take two steps (south then west or west then south) to continue the descent. To compensate for such problem, we perform a minimum search in all neighbors with a topological neighborhood of 2. If a new minimum was found, we continue the gradient descent from it, else the current position is the minimum found by the particle.

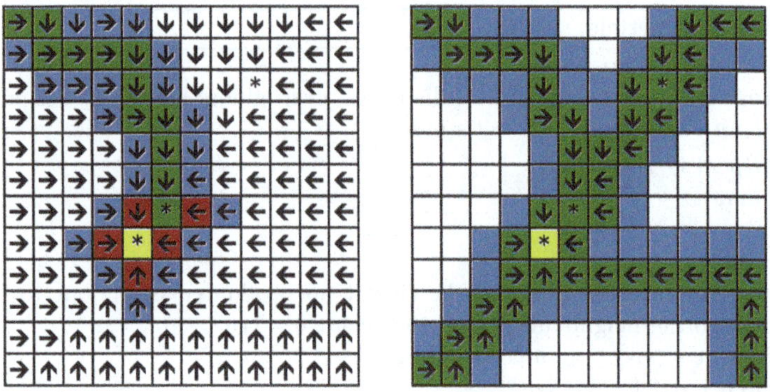

Fig. 1. *Left*: example of a particle execution. Each cell represents a neuron. The arrow in each cell points to the best neighbor (smallest distance). A star represents a local minimum. The green cells are the neurons that were explored during the search, and the blue cells are the neurons whose distance to the input vector was calculated but not selected by the algorithm. After finding a local minimum, new particles are created (in red) and continue the gradient descent by exploring the local neighborhood. The yellow cell is the found minimum, considered as BMU. *Right*: execution of the four particles.

Another problem that can occur with a single *Particle* is edge effects. When projecting high-dimensional data onto a 2D SOM, the map sometimes becomes twisted without a continuous gradient between the upper left and lower right neuron, resulting in particles that may end up in the wrong corner of the SOM, thus finding a very bad BMU so that the update of the codewords may locally break the continuity of the neighborhood. To anticipate such a problem, we start 4 *Particles*, one in each corner of the SOM, and we finally select the smallest BMU found by these *Particles*. We thus preserve the continuity of the SOM and make the Fast-BMU algorithm more robust to local discontinuities in the gradient, especially at the beginning of learning, when the map is not yet well unfolded. The result of a full run of FastBMU is shown in Fig. 1 on the right.

It should also be noted that in the case where there is no clear gradient in the SOM (typically in the first iterations with a randomly initialized SOM), FastBMU does not find the correct BMU. However, it is not a problem for learning, as the neighborhood function rapidly creates a gradient by unfolding the SOM, regardless of the neurons that are initially selected. Once a gradient is created, FastBMU generally finds the correct BMUs.

The next section presents how we adapted the algorithm to a parallel version for future implementations in FPGAs.

4 Parallel Algorithm

One of the properties of the FastBMU algorithm is that it starts from the four corners and converges towards the center during its execution. It means that neurons are used in waves, they do not turn back during the gradient descent. Once an iteration is started, the neurons that are behind the current position of the particle will not be used in the current iteration anymore. Therefore, they are available to handle the next iteration, even before the previous one is completed.

We call *timestep* the time it takes for a particle to evaluate the gradient at its current position and to move to the next neuron. If we launch an iteration of FastBMU per timestep, we use the neurons much more efficiently, without congesting the FPGA. When implemented in hardware the algorithm may thus be able to run an iteration per time it takes to perform a timestep in steady state. Increasing the size of the map does not change the timestep duration, but it only increases the number of iterations simultaneously computed in the map. An illustrative example is shown in Fig. 2.

However, while multiple BMU search in parallel has no impact when performed during the recall phase, since the weights do not change, it modifies the learning phase by possibly modifying the neurons weights during the gradient descent. We measure the impact experimentally in Sects. 5 and 6.

4.1 Neuron Use Rates

In order to have a parallel implementation of FastBMU that is as efficient as possible, the percentage of use for all the neurons in the map has to be as close as possible to 100%, without exceeding it to avoid congestion in the FPGA. By starting one iteration per timestep as presented previously, the corner neurons are used 100% of the time. However, since FastBMU is only following the gradient

Fig. 2. Illustrative example of 3 FastBMU searches performed in parallel, each with a different training vector. If we consider the red one being started at the timestep t=0, then the green one would be launched at t=4 and the blue one at t=7. The image is representing their progress at t=9.

and not evaluating all neurons, other neurons in the map are used significantly less, thereby reducing the global efficiency of our parallel implementation. A typical distribution of neuron use rates (a used neuron has to evaluate its distance to the input) can be found in Fig. 3.

100	100	65	64	63	63	62	59	59	60	60	60	59	60	61	62	63	64	100	100
100	100	65	65	64	64	63	61	61	61	61	61	61	61	63	64	65	65	100	100
56	58	17	18	20	21	20	19	20	20	20	19	18	19	19	19	19	19	61	60
56	58	17	18	20	21	21	20	20	20	19	18	18	19	20	21	19	17	59	57
56	58	17	18	21	22	22	22	21	21	21	20	19	21	21	21	19	18	59	57
56	58	18	19	21	23	23	23	22	22	21	21	21	22	22	21	19	19	59	58
57	59	18	20	22	23	23	24	24	24	23	23	22	23	23	22	20	20	60	58
57	60	19	20	21	23	24	24	24	24	24	24	23	24	22	22	20	19	60	58
58	61	21	20	21	22	23	25	25	25	25	25	24	24	23	21	20	20	60	57
59	61	21	21	22	22	23	24	25	25	25	25	25	24	24	23	20	19	60	58
60	61	19	21	23	23	24	24	24	24	25	25	26	25	24	23	22	20	61	59
60	61	19	19	21	22	24	24	25	25	24	25	25	24	24	23	22	21	62	61
59	61	18	19	21	22	23	24	24	24	25	25	25	24	24	24	23	22	63	61
60	61	19	19	20	20	22	22	23	23	24	24	23	23	23	23	22	22	65	63
62	64	19	19	21	20	21	21	23	23	22	22	22	22	22	22	21	20	66	64
62	64	19	18	20	19	19	20	21	21	20	20	20	20	19	18	19	18	66	64
64	65	19	19	18	17	18	20	20	20	19	18	18	18	18	17	17	16	66	64
65	66	17	17	18	17	18	19	20	20	19	19	18	18	17	16	16	16	66	64
100	100	57	57	57	56	57	58	58	58	57	56	56	56	56	56	56	56	100	100
100	100	55	55	55	54	56	56	56	56	55	54	54	54	54	54	54	55	100	100

Fig. 3. Percentage of evaluations per neuron for FastBMU in a 20 × 20 SOM. The neurons can be divided into three categories. The first one in red is all the corners that are evaluated at each iteration, as the starting point of the particles. The second one is the edges in yellow which are evaluated a little more than half of the time. The last one is the center in green, which are evaluated less than a quarter of the time.

From the figure, we can divide the neurons into three categories, corners, edges and center. The corners are used 100% of the time, the edges between 50 and 60% of the time and less than 25% of the time for the central neurons. Thus, to increase the global neuron use closer to 100%, it is necessary to multiply the use of the central neurons by 4. We can achieve this by increasing the number of iterations started per timestep to 4. However, it also increases the use of the neurons in the edges and the corners beyond 100%. To solve the problem, it is possible in an FPGA implementation to increase the number of comparisons that these neurons can perform per timestep. If we multiply by 4 the number of simultaneous comparisons that we can make for the neurons of each corner for example, then they are able to process 4 iterations simultaneously, and thus they can handle the overload necessary to increase the efficiency of the central neurons. The same principle can be applied for the edge neurons, making them able to perform two or three simultaneous comparisons.

It is important to note that the percentage of use of the central neurons changes with the size of the map. Since FastBMU is more efficient with larger maps, it uses fewer evaluations per iteration, and the percentage of use of the central neurons decreases accordingly. Edge and corner neurons remain at similar percentages of use. Larger maps therefore require more iterations per step if we want to keep central neurons close to the maximum use rate. Therefore the number of possible parallel evaluations by the edge and corner neurons also have to be increased by means of local parallel resources.

By adapting the number of new iterations started per timestep and the number of comparisons possible per neuron, it is therefore possible to get close to a 100% use rate for every neuron of the map. That way, we get the full benefits of the reduced computational cost of the FastBMU algorithm in parallel hardware implementations. The next sections shows the impacts of changing the neurons weights during an iteration of FastBMU in the learning phase.

5 Experimental Setup

We followed a similar experimental setup as in [3], with the same two topologies (Square and Hexagonal) but with only 4 datasets:

1. A uniform distribution in two dimensions in the shape of a square.
2. The pixel color values of an image.
3. A uniformly distributed cube shape that was particularly difficult for FastBMU to learn since a map unfolding in a cube may result in a discontinuous gradient.
4. An image compression application [1] with 10 by 10 sub-images as training vectors (100 pixels with 3 colors each, resulting in 300 dimensions).

The *Parallel* version is a version of FastBMU that launches 4 iterations per step, i.e., there can be several dozens of simultaneous gradient runs at any given time during the learning process. The modification of the weights is done instantaneously after the last particle of a BMU search finishes. This slightly differs from the currently studied hardware implementation, that will be detailed in a future paper. Here, the goal is only to measure the impact of the simultaneity of the iterations on the learning of the SOM with the parallel version of FastBMU.

In order to evaluate the differences between all tested models, we used two metrics: the standard *Mean Squared Quantization Error* (MSQE) which estimates the Vector Quantization quality of the tested algorithm, and the *Mean Squared Distance to Neurons* (MSDtN) which computes the average squared codeword distance between neurons and all of their direct neighbours. The lower the MSDtN value is, the more closely related neighbouring neurons are, and the better the dimensional reduction property is. Numerous metrics exist in the SOM literature to estimate the topographical mapping of the SOM, but MSDtN has the advantage of being easy to compute, without parameters and only dependent on the neurons weights.

$$\mathrm{MSQE} = \frac{1}{V}\sum_{i=1}^{V} ||v_i - w_{\mathrm{bmu}(i)}||^2 \tag{2}$$

$$\mathrm{MSDtN} = \frac{1}{N}\sum_{i=1}^{N}\sum_{j=1}^{N} \begin{cases} ||w_i - w_j||^2, & \text{if } \mathrm{dist}(i,j) = 1 \\ 0, & \text{otherwise} \end{cases} \tag{3}$$

with V the number of vectors in the dataset, N the number of neurons, v_i the weights of the ith vector of the dataset, and $\mathrm{bmu}(i)$ is the index of the neuron which is the best matching unit for the ith vector of the dataset. Similarly, w_i the weights of the ith neuron, and $\mathrm{dist}(i,j)$ is the number of links in the shortest path between neurons i and j in the neural map.

6 Results

The results are presented in Table 1. The first interesting result is for the *Square* data, with a better *MSQE* for the Parallel versions than for the Classical and Fast versions. All the algorithms have a 0% *mismatch* in the recall phase, which means that the difference cannot be explained by the selection of different BMUs in the recall phase, so it probably comes from the simultaneous search of BMUs in the Parallel learning phase. One explanation could be that the simultaneous search adds noise to the learning, which seems to improve the SOM performance in vector quantization but also very to reduce slightly its topological continuity, as shown with the higher *MSDtN* value. The same phenomenon seems to happen for *Colors* data.

With *Cube* data on the other hand, the Parallel learning mode is slightly worse in VQ, but better than the Fast mode when it comes to topological continuity. The Parallel mode also shows the highest percentage of wrong BMUs (*Mismatch*) at 12.8%, when using a grid topology. Finally the three learning modes show similar *MSQE* results with *Image* data.

As first shown in [3], the shape of the data to be learned is the most important factor for the success of the FastBMU algorithm. Our results seem to further

Table 1. Results with a SOM of 32 × 32 neurons, the metrics for each row are computed over an average of 10 runs. The column *Learning* specifies with which algorithm and topology the SOM was trained. MSQE_S is the *Mean Squared Quantization Error* computed with the standard (exhaustive) BMU search algorithm while MSQE_F uses the Fast-BMU version, both during the reconstruction phase. We combine the three different learning modes (Standard, Fast and Parallel) with the two possible reconstruction modes (only Standard and Fast, since the Parallel mode is equivalent to Fast for reconstruction). The differences in MSQE between algorithms thus reflect the quality of the learning phase. The *Mismatch* is the proportion of BMUs that are selected differently by the two algorithms for the reconstruction phase.

Data	Learning	MSDtN	MSQE_S	MSQE_F	Mismatch
Square (2D)	Grid	**2.44e–4**	2.28e–4	2.28e–4	0.0 %
	Grid Fast	**2.44e–4**	2.28e–4	2.28e–4	0.0 %
	Grid Parallel	2.49e–4	**2.02e–4**	**2.02e–4**	0.0 %
	Hexagonal	**2.80e–4**	2.20e–4	2.20e–4	0.0 %
	Hexa Fast	**2.80e–4**	2.20e–4	2.20e–4	0.0 %
	Hexa Parallel	2.84e–4	**2.02e–4**	**2.02e–4**	0.0 %
Colors (3D)	Grid	**1.47e–4**	8.49e–5	8.71e–5	5.3 %
	Grid Fast	1.53e–4	8.63e–5	8.78e–5	5.4 %
	Grid Parallel	1.88e–4	**8.06e–5**	**8.44e–5**	7.6 %
	Hexagonal	**1.62e–4**	8.05e–5	8.10e–5	1.0 %
	Hexa Fast	1.62e–4	7.96e–5	7.98e–5	**0.5 %**
	Hexa Parallel	1.83e–4	**7.60e–5**	**7.68e–5**	1.5 %
Cube (3D)	Grid	5.63e–4	2.22e–3	**2.68e–3**	6.0 %
	Grid Fast	5.80e–4	**2.20e–3**	3.12e–3	7.5 %
	Grid Parallel	**5.59e–4**	2.34e–3	3.77e–3	12.8 %
	Hexagonal	**6.09e–4**	**2.12e–3**	**2.34e–3**	**2.6 %**
	Hexa Fast	6.31e–4	**2.12e–3**	2.51e–3	5.5 %
	Hexa Parallel	6.19e–4	2.18e–3	2.46e–3	3.2 %
Image (300D)	Grid	**2.05e–4**	**1.80e–3**	1.84e–3	3.5 %
	Grid Fast	2.05e–4	1.81e–3	**1.83e–3**	4.5 %
	Grid Parallel	2.23e–4	1.82e–3	1.88e–3	4.7 %
	Hexagonal	**2.34e–4**	1.76e–3	1.78e–3	1.4 %
	Hexa Fast	2.34e–4	1.75e–3	**1.77e–3**	**1.2 %**
	Hexa Parallel	2.49e–4	**1.75e–3**	1.77e–3	**1.2 %**

indicate that on data where the sequential FastBMU works, the parallel version works just as well, and that on data where the sequential version has problems, they get worse with the parallel version.

7 Conclusion

FastBMU is a fundamentally sequential algorithm that speeds up the computation of best matching units in self-organizing maps, taking advantage of the continuity that rapidly appears in the SOM during learning. We propose here an efficient parallel version of FastBMU. Our solution parallelizes the iterations instead of the BMU search as it is made in typical SOM hardware implementations. We have also presented experimental evidence that a SOM learned this way shows comparable performances in vector quantization to the classical SOM, if the shape of the processed data is appropriate.

The described work paves the way to an efficient implementation of FastBMU in FPGA systems, thereby proposing a fast and completely cellular approach to self-organizing maps. The hardware implementation work is in progress, with an ongoing study of how to efficiently find the BMU between the four particles in an FPGA without using too much memory or communications.

References

1. Amerijckx, C., Legat, J.D., Verleysen, M.: Image compression using self-organizing maps. Syst. Anal. Model. Simul. **43**(11), 1529–1543 (2003)
2. Astudillo, C.A., Oommen, B.J.: Fast BMU search in SOMs using random hyperplane trees. In: Pham, D.-N., Park, S.-B. (eds.) PRICAI 2014. LNCS (LNAI), vol. 8862, pp. 39–51. Springer, Cham (2014). https://doi.org/10.1007/978-3-319-13560-1_4
3. Bernard, Y., Hueber, N., Girau, B.: A fast algorithm to find best matching units in self-organizing maps. In: Farkaš, I., Masulli, P., Wermter, S. (eds.) ICANN 2020. LNCS, vol. 12397, pp. 825–837. Springer, Cham (2020). https://doi.org/10.1007/978-3-030-61616-8_66
4. Fritzke, B.: A growing neural gas network learns topologies. In: Advances in Neural Information Processing Systems, vol. 7, pp. 625–632. MIT Press (1995)
5. Jovanović, S., Hikawa, H.: A survey of hardware self-organizing maps. In: IEEE Transactions on Neural Networks and Learning Systems (2022)
6. Kohonen, T.: Self-organized formation of topologically correct feature maps. Biol. Cybern. **43**(1), 59–69 (1982)
7. Kohonen, T.: Essentials of the self-organizing map. Neural Netw. **37**, 52–65 (2013)
8. MacQueen, J.: Some methods for classification and analysis of multivariate observations. In: Berkeley Symposium on Mathematical Statistics and Probability, pp. 281–297 (1967)
9. Marsland, S., Shapiro, J., Nehmzow, U.: A self-organising network that grows when required. Neural Netw.: Official J. Int. Neural Netw. Soc. **15**(8–9), 1041–1058 (2002)
10. Martinetz, T.M., Berkovich, S.G., Schulten, K.J.: Neural-gas network for vector quantization and its application to time-series prediction. IEEE Trans. Neural Netw. **4**(4), 558–569 (1993)
11. Yin, H.: The self-organizing maps: background, theories, extensions and applications. In: Kacprzyk, J., Fulcher, J., Jain, L. (eds.) Computational Intelligence: A Compendium. Studies in Computational Intelligence, vol. 115. Springer (2008). https://doi.org/10.1007/978-3-540-78293-3_17

Neural Networks for Spatial Models

Cécile Hardouin[1] and Jean-Charles Lamirel[2(✉)]

[1] Modal'x, University Paris Nanterre, Nanterre, France
hardouin@parisnanterre.fr
[2] Synalp, LORIA, Vandoeuvre-les-Nancy, France
lamirel@loria.fr
http://hardouinc.perso.math.cnrs.fr/

Abstract. The aim of spatial econometrics is to analyze and/or predict the relationship between one dependent variable Y with other variables, building a model that takes into account the spatial dependence. Usual spatial econometric models are based on a neighbourhood matrix whose elements are linked to geographical distances. We propose to use distances between prototypes resulting from a neural classification instead. The results are at least as well as the ones obtained from the geographical distances based design. In some cases, we need two neighbourhood matrices and the difficulty rising then is to find a second matrix; then this issue is simply solved by using one matrix based on geographical distances and the other based on neural distances. Finally, the use of neural distances opens the door to use spatial econometrics methods for the analysis and the modelling of any data sets, non necessary geographically referenced. We illustrate our approach with two studies of real datasets.

Keywords: SOM · Growing Neural Networks · Clustering · Spatial statistics

1 Introduction

Nowadays, many data sets include geographical contents. Spatial data occur in various domains; in ecology or epidemiology, in economics and social sciences, or image analysis. There are many cases where the geographical aspects ensure spatial dependence, when the data are collected from neighbouring locations. Indeed, it is frequently observed that sample values recorded on one location tend to be similar to those collected from nearby locations. Therefore, this dependence can't just be ignored. Usual statistical models do not apply in the case of correlated data; moreover, spatial dependence needs a specific approach. There are several ways to account for spatial interaction; spatial econometrics propose regression models involving spatial neighbourhood matrices induced by neighbourhood graphs. This neighbourhood matrix is a key element in the model design. A frequent framework is to consider weights linked to the geographical

J. Faigl et al. (Eds.): WSOM+ 2022, LNNS 533, pp. 21–30, 2022.
https://doi.org/10.1007/978-3-031-15444-7_3

distances between the locations where the data were collected from. The question we address here is to build neighbourhood weights based on the specific aspects of the data, using clustering based on neural network. This method can also be used for any data where observations are collected on a graph structure, the latter playing the role of a neighbourhood graph. Moreover, we show that we can use this technique for any data set.

We consider the class of Spatial Autoregressive Models where the dependent variable is linked to other variables (covariates) and to the values of its own adjacent values. The design needs to determine what means to be "adjacent" through a neighbourhood graph associated to a neighbourhood matrix W; this matrix $W = (w_{ij})$ quantifies the connections between individual locations, filled with zeros for non adjacent sites and positive weights for neighbouring sites. Usually the weights w_{ij} are linked to the geographical distance d_{ij} between neighbouring sites i and j, they are proportional to the distance at a negative power $d_{ij}^{-\alpha}$, see [7]. Our purpose in this work is to propose distances based on neural clustering. Indeed, similar values of the dependent variable can result from geographical proximity, but also from similarity of the covariates. We need them to define other measures of similarity. Our approach is the following. First, we run a neural algorithm on the covariates in order to obtain clusters, like a Kohonen SOM algorithm or a growing neural gas algorithm (GNG). This procedure also provides distances between nodes. We then use directly these distances for the weights of the neighbouring matrix W; or we combine the geographical distances and the clustering distances.

This approach can be generalized to non geographical data. A fist trivial extension is to consider any data on a graph structure, considering network autocorrelation rather than spatial autocorrelation, see [11], but more generally our approach can be considered for any data set. We adopt the general concept of a neighbourhood matrix based on distances between prototypes after neural clustering. This provides a full data-driven approach very useful for analysis or prediction. It is important to run a clustering, because sparse matrices W, or decaying weights, give better results, see for instance

Section 1 describes the two algorithms we use for clustering. We experiment both Kohonen maps and Growing Neural Gas networks. We additionally shortly describe the method used to select the best number of clusters. We recall the definition of some common spatial econometrics models in Sect. 2. Then we illustrate our approach on real data sets in Sect. 3; in the first example we consider indicators of the intensity of the COVID-19 epidemic (up to May 2020) recorded in departments in Metropolitan France; some socio-economic variables are also recorded and are used as covariates in the models. We run several models using the same neighbourhood graph, but with geographically-based distances and/or neural-based distances neighbourhood matrices. The results show evidence of socioeconomic factors in the Covid-19 occurrences, but the use of neural clusters distances improves the results. The second example is a data set without spatial coordinates; we build a spatial weight matrix based on clustering on the covariates; the results show that spatial models using this weight matrix provide better results than ordinary regression.

2 Neural Clustering and Optimal Clustering Model Detection

2.1 Neural Clustering

The neural clustering methods are based on the principles of neighbourhood relation between clusters, either they are preset (fixed topology), like the "Self-Organizing Maps" also named SOM [8], or dynamic (free topology), like static "Neural Gas" (NG) [13] or "Growing Neural Gas" (GNG) [6]. As compared to usual clustering method like K-means [12], this strategy makes them less sensitive to the initial conditions, which represents an important asset in many data analysis tasks.

The architecture of the SOM network is based on the understanding that the representation of data features might assume the form of a self-organizing feature map which is geometrically organized as a grid. A mapping from a high dimensional data space $\mathbb{R}^n$ onto a two dimensional grid of neurons is thus defined. The SOM algorithm is presented in details in [8]. It consists of two basic procedures: (1) selecting a winning neuron on the grid and (2) updating weights of the winning neuron and of its neighbouring neurons. Once the SOM algorithm is achieved, the training data can be affected to their nearest neuron of the map.

In the NG algorithm [13], the weights of the neurons are adapted without any fixed topological arrangement within the neural network. Instead, this algorithm utilizes a neighbourhood ranking process of the neuron weights for a given input data. The weight changes are not determined by the relative distances between the neuron within a topologically pre-structured lattice, but by the relative distance between the neurons within the input space, hence the name "neural gas" network. Indeed, thanks to the loss of topographic constraints as compared to SOM, NG tends to better represent the structure of the data, theoretically yielding to better classification results. Nevertheless, one of the main drawbacks of the NG algorithm is that the number of neurons, materializing the final number of clusters, represents a fixed parameter.

The GNG algorithm [6] solves the static character of the NG algorithm bringing out the concept of evolving network. The number of neurons is adapted during the learning phase according to the characteristics of the data distribution. Indeed, GNG gives the possibility of creating and removing neurons, as well as connections between them. The suppression process relies on the notion of maximal age, or maturity, of a connection. At time t, a connection reaching its maturity is automatically deleted. Isolated neurons are consequently removed. On the other hand, the creation of the neurons is made only periodically (each T iteration or time period) between the two neighbouring neurons that accumulated the most important error for representing the data. Different stopping criteria can be used like a maximal number of neurons or a minimal change between the reference vectors of the neurons between two time periods.

2.2 Optimal Clustering Model Detection

Optimal clustering model detection is still an open research problem although its main importance in all clustering tasks. For that purpose, Dugué and al. [5] show the interest of discarding the use of classical distances or minimizing square-error based approaches, which strongly depend on the clustering method and work badly with high dimensional and sparse representations, and with non Gaussian clusters as well. They propose an alternative approach that relies on analysis of the features of clusters' related data and thus presents the main advantage of being totally independent of the clustering algorithm. Selected features for identifying optimal clustering model are the ones that are both the most discriminant and the most general for describing clusters' content. This results in highlighting the clustering model that optimizes the contrast based on the said features. This approach substitutes to usual compactness and separation measures. More details on said approach can be found in [5]. This approach is resistant to noise and outperforms all existing approaches for optimal clustering model detection on real data, whatever low dimensional or high dimensional data are considered.

3 Simultaneous Autoregressive Models

3.1 Spatial Process

We consider a spatial process defined on a set of sites $S \subset \mathbb{R}^d$. These sites can be nodes on a regular or irregular grid, pixels, or centroids of small geographical areas. In our first example, S is finite and our sites are centroids of French departments in metropolitan France. We observe a process Y, with some covariates $X_1, \cdots, X_p$, on locations $\mathbf{s}_1, \ldots, \mathbf{s}_n$. The goal is to model Y as a function of $X_1, \cdots, X_p$.

Spatial autocorrelation is defined as the correlation of a process $\{Y(\mathbf{s}),\ \mathbf{s} \in S\}$ with himself. It measures the degree of similarity between neighbour observations; if it is positive, there's cooperation, neighbouring sites tend to have similar values; if it is negative, there's repulsion, neighbouring sites tend to have very different values. Spatial autocorrelation can be inherent to the process, or results from outside effects, or even remains unexplained by the variables at disposal.

3.2 Neighbourhood

To measure spatial autocorrelation, we first have to define the neighbourhood, that is to define a neighbourhood graph. The latter is without loop and can be symmetric or not. Typically, it is based on the geographic arrangement of the observations, or simple contiguity. In the case of regular lattices, the graph of the four nearest neighbours is the most prevalent. In the case of irregular areas, it is common to consider as neighbours areas sharing a common border or within a given distance of each other. Thus, it is very common to use the so-called Queen contiguity; it defines neighbours as two polygons that share a common boundary or vertex. Note that this neighbourhood is symmetric.

Once the neighbourhood graph is defined, we turn to the neighbourhood or also called weights matrix W. For each location i, $w_{ii} = 0$, $w_{ij} = 0$ if j is not a neighbour of i, and $w_{ij} > 0$ otherwise. So, the spatial weights matrix depends crucially on the graph. The choice of the non-zero weights is again a key step in the model as we see later.

3.3 Spatial Econometrics Models

We present hereafter some of the most common spatial econometric models used in spatial regression models. All of them belong to the class of simultaneous auto-regressive models. See for instance [3,9] for a complete review.

In its simplest form, a spatial auto-regressive process is described by the equation $Y = \rho WY + u$, where W is the spatial weights matrix, WY is the spatially lagged dependent variable, ρ is the auto-regressive parameter, and u is the error term. It is well defined as soon as $(I - \rho W)$ is not singular. If u is a white noise, then this refers to the *spatial lag model*, analogously to time series, W being the lag operator.

The *mixed spatial auto-regressive model* (SAR) takes the following form,

$$Y = \rho WY + X\beta + \varepsilon.$$

The *spatial error model* (SEM) is defined by

$$Y = X\beta + u, u = \lambda Wu + \varepsilon.$$

The *general spatial model* (SAC) is

$$Y = \rho W_1 Y + X\beta + u, \quad u = \lambda W_2 u + \varepsilon.$$

In these writings, Y is the dependent variable, X is a matrix of exogenous variables, W is the spatial weights matrix, and ε is the error term, usually $\varepsilon \sim N(0, \sigma^2 I)$. Parameters ρ and λ account for the spatial dependence strength. In the mixed spatial auto-regressive model, it is assumed that Y's values observed on different locations depend on one-another directly as a linear combination of neighbouring locations values with other explanatory variables. The spatial error model is chosen when the spatial dependence seems to result from unobserved processes. Finally, one might consider the general spatial model when either the spatial lag model or the spatial error model fail. It is recommended then to consider different weight matrices W_1 and W_2. See for instance [2] for the estimation of these models.

3.4 Spatial Weights

All the models above involve weights matrix W (or W_1 and W_2). As seen before, the elements of this matrix are non stochastic, and their choice is crucial. Binary matrices reflecting contiguity are often dropped in the case of irregularly spaced locations; it is common to consider $w_{ij} = d_{ij}^{-\alpha}$ for neighbour sites i and j, where $\alpha > 0$ and d_{ij} is the geographical distance between i and j. Let us note that in this case, the geographical effect is both in the neighbourhood construction and in the weight itself.

We propose here to use another index of vicinity. Let us consider the SAR and the SEM models, with the same neighbourhood graph. Let us assume that this neighbourhood has been chosen on geographic reasons; let us consider the Queen neighbourhood without loss of generality. Then, two neighbour sites are geographically closer one to the other than two sites chosen at random. In the meantime, we can also suspect spatial dependence in the covariates; the Durbin spatial model is tempting but is sensitive to colinearity troubles. Our approach overcomes this issue, and allows to consider other distances based on similarity between covariates values. Moreover, when the SAR or the SEM models do not provide sufficiently good results, one is interested to run the general spatial model, but is frequently stopped by the choice of the second weights matrix W_2. Then our approach makes it possible by setting W_1 as a geographical distances-based matrix and W_2 as a neural distances-based matrix. We explain below what we intend by neural distances.

The idea is to run a clustering algorithm on the covariates $X_1, \cdots, X_p$. If we run a SOM algorithm, we obtain a self-organized map; let us note that this method preserves topology, in the sense that similar input values are associated to the same cluster or are located in neighbouring clusters on the output map. We also consider topology-free neural clustering based on growing gas (GNG), as explained in Sect. 2. After the run of one or the other algorithm, each multivariate observation is associated to a prototype and we store this information together with the distances between the prototypes. This means that each site i has now two kinds of neighbours: geographic neighbours and neural neighbours. Let us denote now d_G the geographical distances and d_N the neural distances. Our purpose is to use distances d_N into the spatial weight matrix, whose non-null elements are now $w_{ij} = d_{N,ij}^{-\alpha}$. We illustrate the method on real data sets in the following section.

4 Application

We consider a real data set about the Covid-19 epidemic in Metropolitan France in 2020. The unit areas are 96 french departments and the set of sites S comprises the centroids of these 96 departments. The dependent variables that we consider are some indicators of the epidemic, that is the *rate of hospitalization* due to the disease, the *rate of death* at hospital due to the disease, and the *rate of mortality excess*. The covariates are some socioeconomic variables: the *density of population*, the *proportion of the three most important cities* in the department's population, the *rate of second home* (as a part of all housing), the *rate of inactive people* (excluding retirements), the *rate of workers*, the *rates of emergency service* and of *physician* (relatively to the population). These data set have been studied by ([1], 2020) who highlighted the role of socioeconomic covariates in the expansion of the disease. For technical reasons, some of the variables are considered after a log-transformation.

Figure 1 highlights the presence of spatial autocorrelation with the following map of the log of death at hospital. Similarly, the maps of mortality excess or log of hospitalization rate display obvious spatial correlation.

For this study, the Queen neighbourhood has been considered, as illustrated in Fig. 1.

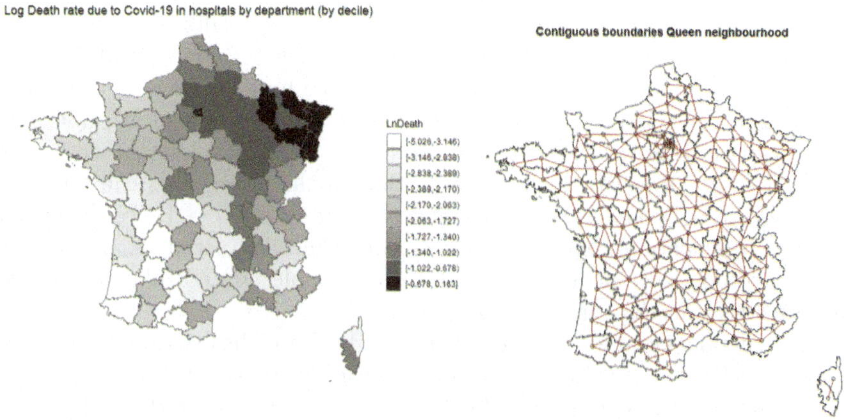

Fig. 1. Left: deciles distribution of the log of the *death rate*. Right: the Queen neighbourhood

First, we run the two neural clustering algorithms (i.e., SOM and GNG) on the seven covariates. For each method, the number of clusters has been determined following the procedure described in Sect. 2. As a result, the optimal Kohonen map is an hexagonal map of size 4 by 3 and the optimal GNG gas involves 12 clusters. Then, we build weight matrices W_G, W_K and W_{GNG} defined by respectively geographical distances, Kohonen distances, growing neural gas distances (distances between the prototypes for the two latters). We used inverse distances and square root of inverse distances; each matrix is normalized by its larger eigenvalue (normalizing the weight matrix is standard in spatial regression and this kind of normalization allows comparisons). Then we run the spatial regression models. Though the clustering algorithms are stochastic methods, the regressions results are very robust.

Let us summarize the results that we obtained. Table 1 below presents the detailed results for *death excess*; we skip the ones for the other variables for sake of place. Since we are more interested in this work by variables selection, we present the estimate value of the regression coefficient together with its p-value, rather than the usual standard error. One important coefficient is ρ or λ which parametrizes the influence of the neighbouring values. This coefficient must be significant. Finally, the table displays the AIC values which helps to select the best model. Let us recall that when the parameter ρ or λ is significantly not null, then the interpretation of the regression coefficients is different from a conventional least squares interpretation, see for instance [11]. We consider *impacts* instead. Since our goal is to present a model approach rather than studying the socio-economic effects on the virus contamination, we won't go into details on the interpretation, but just look on sign and significance.

The models select the same covariates as being significant for the dependent variable, but the best AIC is often obtained for the W_K weights. The fact that the models give similar results is not so surprising, according to [10]; but it is nice to confirm that clustering distances provide good results; it means that, in such a way, we can use them to derive spatial models for non spatial data.

For the *hospitalization* dependent variable, the best model is a SAR model, with weight matrix $W_{1/\sqrt{d_K}}$; it is the same for the dependent variable *death excess*; for the *death* variable, the best model is a SEM model, again with weight matrix $W_{1/\sqrt{d_K}}$. It is worthwhile noting that in the three cases the significant covariates are the *population density* and the *rate of workers*. However, we note that the coefficients of *emergency service* and *physician* variables are negative, which corresponds to intuition, though these variables are not playing the most important roles in the spread of the disease.

The general spatial models were ran using both weight matrices $W_{1/\sqrt{d_G}}$ and $W_{1/\sqrt{d_K}}$ (or $W_{1/\sqrt{d_{GNG}}}$); in the cases of the *hospitalization rate* and *death excess*, the results reject the significance of parameter λ, which confirm the SAR hypothesis; while for the *death* variable, it is the parameter ρ which is not significant, and confirms that the right model seems to be a SEM model.

The second example illustrates how we can use spatial models for non geographical data. Our data set contains information about 474 employees hired by a Midwestern bank between 1969 and 1971. The file contains beginning salary (`salbegin`), salary now (`salary`), seniority (`jobtime`), previous experience (`prevexp`), education level (`educ`) gender (`sex`), and minority belonging (`minority`). The dependent variable is the salary, and for better results, this variable, as well as `salbegin` will be log-transformed. We run a clustering on the quantitative covariates; then we build a weight matrix W using the inverse distances between the prototypes. Here an optimal 4 by 3 Kohonen map and an optimal 9 clusters GNN gas were used. We compare ordinary regression model with spatial models. Results are summarized in Table 2. SEM models are not relevant for this data set, but SAR models improve the AIC of the classic regression model. We did run several times the neural algorithms, the results are very steady and always give better AIC than OLS procedure.

Table 1. SAR Regression coefficients estimates for Death excess, p-value significance in parenthesis

Model	SAR $W_{1/\sqrt{d_G}}$	SAR $W_{1/\sqrt{d_K}}$	SAR $W_{1/\sqrt{d_{GNG}}}$
(Intercept)	−0.544 (0.03042)	−0.559 (0.01982)	−0.535 (0.02651)
Ln Pop Density	0.052 (0.02418)	0.077 (0.00016)	0.086 (2. 10^{-5}
Second Home	0.001 (0.59989)	0.002 (0.29267)	0.002 (0.25308)
Inactive	1.044 (0.25850)	0.490 (0.57568)	0.346 (0.69484)
Work	2.966 (0.00480)	2.494 (0.01169)	2.235 (0.02463)
Emergency	−0.002 (0.56082)	−0.002 (0.62424)	−0.001 (0.72948)
Physician	−0.087 (0.45960)	−0.090 (0.42451)	−0.118 (0.29517)
ρ	0.8092 (5. 10^{-11})	0.7326 (2.10^{-12}	0.70972 (4. 10^{-11}
AIC	−80.581	−86.948	−85.499

Table 2. Regression results for OLS and SAR models. Coefficients: estimate, standard error and p-value in parenthesis. ρ estimate with p-value in parenthesis. AIC.

Model	OLS	SAR W_{1/d_K}	SAR $W_{1/d_{GNG}}$
Intercept	1.226	2.0963	1.9788
	0.3204 (0.000147)	0.4387(1.771.10^{-6})	0.40920 (1.327. 10^{-6})
logsalbegin	0.8996	0.83538	0.84328
	0.03457 (< 2.10^{-16})	0.04078 (< 2.10^{-16})	0.039206 (< 2.10^{-16})
jobtime	0.004541	0.00507454	0.0048986
	0.0007570 (3.98.10^{-9})	0.00077 (3.959.10^{-11})	0.00075526 (8.818.10^{-11})
prevexp	−0.0005373	−0.00072492	−0.00070847
	7.986e−05 (5.03.10^{-11})	0.00010254 (1.556.10^{-12})	9.8551e−05 (6.533. 10^{-13})
educ	0.01091	0.00981151	0.011262
	0.003909 (0.005459)	0.00386719 (0.01118)	0.0038484 (0.0034282)
minorityYes	−0.02881	−0.02671323	−0.032103
	0.01901 (0.130276)	0.01872284 (0.15365)	0.018739 (0.0866863)
sexFem	−0.05509	−0.06005374	−0.065571
	0.01873 (0.003430)	0.01851372 (0.00118)	0.018782 (0.0004811)
ρ		−0.025688 (0.0045196)	−0.021912 (0.0040875)
AIC	−361.84	−367.91	−368.09

5 Conclusion

In this work, we present an interesting and new regression model, using neural methods and spatial statistics. We show that for spatial data, the use of neural distances obtained from a preliminary clustering on the covariates allows to derive similar results than traditional geographical distances, and often provides better accuracy in the case of the SOM algorithm. This new weight matrix based on neural distances between cluster nodes is to be considered as well as geographical distance-based weights for a SAR or SEM model, but it is a natural choice for the "second" weight matrix in general spatial models; one could even think at this matrix to be the 'first" matrix W_1 and take the geographical-based matrix as W_2.

Another important feature of this work is to provide a regression method for non spatial data. We can model any regression model as a spatial regression model. We present on an example how our modelling procedure can improve the results. In the latter example, clusters have been obtained from the quantitative covariates only; of course qualitative variables should be taken into account and this is a work in progress.

Notes and Comments. This research has been conducted as part of the project Labex MME-DII (ANR11-LBX-0023-01).

References

1. Amdaoud, M., Arcuri, G., Levratto, N.: Covid-19: analyse spatiale de l'influence des facteurs socio-économiques sur la prévalence et les conséquences de l'épidémie dans les départements français. No 4, EconomiX Working Papers from University of Paris Nanterre (2020)
2. Anselin, L.: Spatial Econometrics: Methods and Models. Kluwer, Dordrecht (1988)
3. Anselin, L., Bera, A.-K.: Spatial dependence in linear regression models with an introduction to spatial econometrics. In: Ullah, A., Giles, D.E.A. (eds.) Handbook of Applied Economic Statistics, pp. 237–289. Marcel Dekker, New York (1998)
4. Bivand, R.S., Pebesma, E.J., Gómez-Rubio, V.: Applied Spatial Data Analysis with R Use R!, Secaucus. Springer, New YorK (2008). https://doi.org/10.1007/978-1-4614-7618-4
5. Dugué, N., Lamirel, J.-C., Chen, Y.: Evaluating clustering quality using features salience: a promising approach. Neural Comput. Appl. **33**(19), 12939–12956 (2021). https://doi.org/10.1007/s00521-021-05942-7
6. Frizke, B.: A growing neural gas network learns topologies. Adv. Neural. Inf. Process. Syst. **7**, 625–632 (1995)
7. Getis, A., Aldstadt, J.: On the specification of the spatial weights matrix. Geogr. Anal. **35** (2004)
8. Kohonen, T.: Self-organized formation of topologically correct feature maps. Biol. Cybern. **43**, 56–59 (1982). https://doi.org/10.1007/BF00337288
9. LeSage, J., Pace, R.K.: Introduction to Spatial Econometrics. Chapman and Hall/CRC Press, London (2009)
10. LeSage, J.P., Pace, R.K.: The Biggest Myth in Spatial Econometrics (2010). https://doi.org/10.2139/ssrn.1725503
11. LeSage, J.P.: An Introduction to Spatial Econometrics. Revue d'économie industrielle 123 (2008)
12. MacQueen, J.: Some methods of classification and analysis of multivariate observations. In: Proceedings of the 5th Berkeley Symposium in Mathematics, Statistics and Probability, pp. 1:281–297, University of California, Berkeley, USA (1967)
13. Martinetz, T., Schulten, K.: A Neural Gas Network Learns Topologies, Articial Neural Networks, pp. 397–402 (1991)
14. Pace, R.-K., Lesage, J.P.: Interpreting Spatial Econometric Models, the Regional Science Association International North American Meetings, Canada, Toronto (2006)

Machine Learning and Data-Driven Approaches in Spatial Statistics: A Case Study of Housing Price Estimation

Sarah Soleiman[1,2(✉)], Julien Randon-Furling[1,3], and Marie Cottrell[1]

[1] SAMM, Université Paris Panthéon-Sorbonne – FP2M (FR2036) CNRS, Paris, France
sarah.soleiman@gmail.com
[2] MeilleursAgents, 7 Boulevard Haussmann, Paris, France
[3] Department of Mathematics, Columbia University, New York, USA

Abstract. The intertwining of socio-spatial complexity with that of price formation leads to highly challenging questions when modeling real estate markets and the dynamics of property prices. The exact same apartment typically will not have the same price depending on its location in the city - due to specifics of the neighborhoods and even micro-neighborhoods that are difficult to quantify. Traditional methods rely on the so-called hedonic approaches modified to incorporate spatial effects via geographically weighted regressions. However, the recent availability of big data pertaining to the socio-economic characteristics of cities, at a very fine-grained level, should allow one to capture in much finer detail the complex relationship between space and price in the real estate market. Our approach is two-fold, we first apply a simple Self-Organizing Map (Kohonen) algorithm on vast sets of demographical, economical and infrastructural data in order to bring out the socio-spatial structure of a city and then use this cluster information into the spatial diffusion process of the GWR. SOM gives a notion of proximity between clusters and thus provides a multi-scale degree of similarity (unlike other clustering algorithms like k-mean).

Keywords: SOM · Spatial statistics · Real estate dynamics

1 Indroduction

Spatial modeling of the real estate prices is justified by the idea that prices are highly correlated within geographical areas. These issues, related to the valuation of real estate properties and the identification of real estate sub-markets, have been approached from different angles, from so-called hedonic models or more complex geo-statistical models [4,8–10].

One of the most widely used models is the Geographically Weighted Regression (GWR). In particular, Bitter et al., in 2006, study the phenomenon of spatial heterogeneity in Arizona using the spatial expansion method and GWR [1]. The

J. Faigl et al. (Eds.): WSOM+ 2022, LNNS 533, pp. 31–40, 2022.
https://doi.org/10.1007/978-3-031-15444-7_4

application of geographically weighted regression gives much better results than the spatial expansion method.

However, modern urban areas are very heterogeneous in nature, comprised of residential blocks, metro lines, parks and other geographical landmarks. Each of these structures interfere with real estate prices in its neighborhood. These urban areas are also characterized by strong socio-spatial dissimilarities.

In this study, we describe socio-spatial patterns present in residential areas using public socio-economic data and incorporate this piece of information into real estate prices model.

2 Data

We used databases from Meilleurs Agents. Raw data are apartment's transactions described by the date (yyyy-mm-dd) of the transaction, it is exact location (land plot level), it is price and characteristics describing the apartment: the number of rooms, floor, area, price, presence/absence of an elevator. We use 5 years of those past transactions, between january 2014 and september 2019 in the city of Les Lilas only, which is a total of $N = 386$ past transactions.

Pre-processing of the data is two-fold: we first apply filters to avoid atypical apartments and get rid of the outliers (number of rooms must not exceed 15, area between 8 and 500 m^2), and then update the price of transactions adjusted for the real estate price index. The latter step enables to compare transactions at a fix time, we choose to set the model date at 2019-09-01 (Table 1).

Table 1. Description of the training set

Label	Variable	Min	1st quartile	Median	3th quartile	Max	Total
room_count	number of rooms	1	2	2	3	6	386
floor	floor	0	1	2	4	18	386
area	area (in m^2)	13	36	47	68	170	386
price	price (in €)	24 000	200 000	281 000	400 000	945 000	386

47% of the transactions have a presence of an elevator in their description.

3 Spatial Diffusion Process

We apply GWR (Geographically Weighted Regression) [6], as a spatial diffusion model on housing transaction prices. GWR consists essentially in a classical regression where observations are weighted according to geographical distance to the location of the point considered.

To explain the GWR model, we follow the presentation done by M. Charlton in [6]. For each transaction i, $i = 1, ..., N$, one knows its price P_i (dependent variable) and p independent variables $x_i^1, ..., x_i^p$, as well as its position in a geographical system. For the sake of simplicity, we use i to denote the transaction, the index of the transaction and its location itself.

The main equation related to a transaction at location u can be written as

$$P_i(u) = \beta_{0i}(u) + \beta_{1i}(u)x_i^1 + ... + \beta_{pi}(u)x_i^p + \varepsilon_i. \tag{1}$$

Unlike a classical linear regression model, parameters $\beta_{0i}(u), \beta_{1i}(u), ..., \beta_{pi}(u)$ depend on location u.

Equation (1) represents a weighted regression model where the function to be minimized is

$$\mathcal{E}(u) = \sum_{i=1}^{N} w_i(u)\varepsilon_i^2, \tag{2}$$

where

$$w_i(u) = \exp \frac{||i-u||^2}{2\sigma^2} \tag{3}$$

is the geographical weight, which takes into account the distance (in meter) between location u and locations of other transactions. Parameter σ is chosen by cross-validation. In the following, $w_i(u)$ is denoted by $w_i^{GEO}(u)$.

The estimation of vector $\beta(u)$ is

$$\hat{\beta}(u) = (X^T W(u) X)^{-1} X^T W(u) P(u), \tag{4}$$

where $W(u) = diag(w_1(u), ..., w_N(u))$, X is the independent variables ($N \times (p+1)$) matrix (whose first column is only composed of 1) and P is the N-vector of prices.

However, two apartments located on both sides of the same street can present very different housing quality. Thus, considering the geographical distance only would bias the estimation. To account for this, we decide to use other variables that would describe the neighborhood atmosphere. We use public data from the French national office of statistics - INSEE (Institut national de la statistique et des études économiques). A clustering of location based on these extra variables will allow us to better qualify the apartments in the studied area.

4 GWR × SOM: Merging Socio-spatial Information into a Spatial Diffusion Process

4.1 SOM

We choose to work with SOM as a clustering algorithm because it preserves topology: neighboring (i.e., similar) observations in the input space are located on the same or neighboring clusters on the SOM map.

HAC is then applied on prototypes to produce super-clusters that are useful to visualize and interpret the socio-spatial structure of a city on a simple geographical map.

The data bases are provided on a grid of 200 × 200 m cells covering the entire country. Since cell division does not take into account geographical, natural or urban delimitations, we map cell data to the block level, weighting by surface overlap. The city of Les Lilas is composed of 67 blocks in total. The set of $q = 13$ variables we use includes socio-economic data such as age and income distribution, percentage of household owners, percentage of apartments in a block, etc.

Let us denote the number of blocks by n and $y_j \in R^q$ (space R^q is endowed with the Euclidean distance) the feature vector of block j for $j = 1, ..., n$. We run SOM on all y and get clusters $C_1, ..., C_K$ represented by prototypes $m_1, ..., m_K$. Vector y is assigned to a cluster by

$$y \in C_{k_0} \Leftrightarrow ||y - m_{k_0}|| = \min_{k=1,\dots,K} ||y - m_k||. \tag{5}$$

We denote by $C_{k_0(y)}$ the cluster y belongs to and $m_{k_0(y)}$ its prototype vector. The SOM distance between vectors y and y' is defined by

$$d_{SOM}(y, y') = ||m_{k_0(y)} - m_{k_0(y')}||. \tag{6}$$

For any transaction i, we are looking for the block j that contains it, denoted by $j(i)$ and the extra vector y associated to i will be the vector $y_{j(i)}$.

We define SOM weights as

$$w_i^{SOM}(u) = \exp\left(-\frac{d_{SOM}(y_{j(i)}, y_{j(u)})^2}{2\gamma^2}\right) \tag{7}$$

with γ a non-negative parameter chosen by cross validation. In the present work, for each transaction u, we only take into account the transactions i such that $y_{j(i)}$ and $y_{j(u)}$ belong to neighboring clusters on the SOM map for a radius 1 (same cluster or adjacent ones).

4.2 Final Model

Hence, final weights are defined by

$$w_i^F(u) = w_i^{GEO}(u) \times w_i^{SOM}(u). \tag{8}$$

The function to be minimized is now

$$\mathcal{E}(u) = \sum_{i=1}^{N} w_i^F(u)\varepsilon_i^2. \tag{9}$$

If we consider a neighboring point i that is on the exact same location of the current point u, then $w_i^{GEO}(u) = 1$. And if i and u belong to the same SOM cluster then $w_i^{SOM}(u) = 1$. In this case $w_i^F(u) = 1$. These weights decrease to 0 when i is far from u in a geographical sense and/or in a socio-economic sense. An example of the impact of the new model on weights is shown on Fig. 1c for a given location.

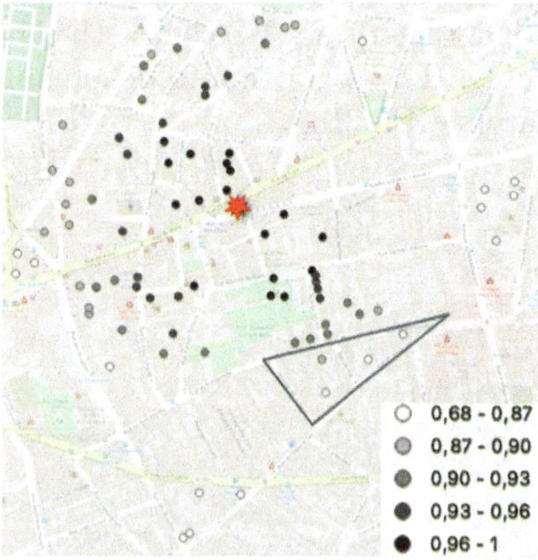

(a) Geographical weights for the considered point (red star).

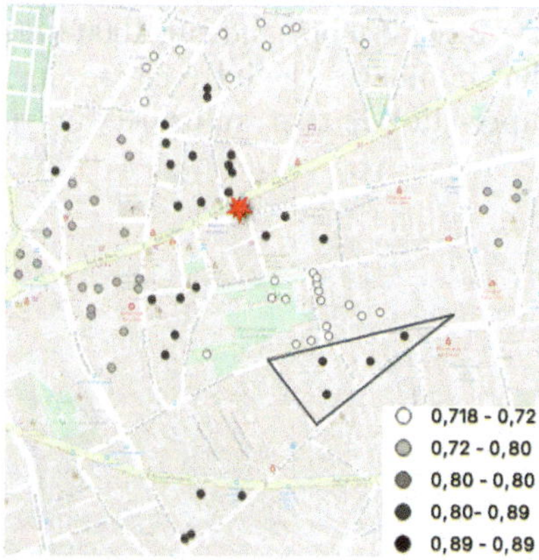

(b) SOM weights for the considered point (red star).

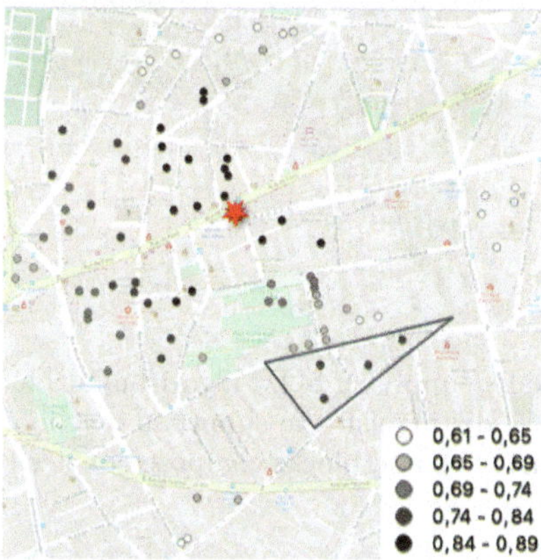

(c) Final weights for the considered point (red star).

Fig. 1. Geographical, SOM, and final weights. If we have a closer look at weights inside the black triangle, transactions located in this area will have more importance with our new model (c) than if we had only taken into account geographical distances (a). Indeed, neighborhood of this area and the one of the considered point (red star) appears to be similar (b).

5 Results

As an example, we show here the results obtained on the city of Les Lilas, comprising 43 rectangles and 67 blocks – a small city just outside central Paris, at the heart of the 12-million Île de France metropolitan region. We set 9 clusters (3×3 SOM map) and run 1000 iterations of SOM algorithm. We choose the one that minimizes the intra-classes inertia. We use a HAC algorithm to define four super-clusters that we project on the city map (see Fig. 2).

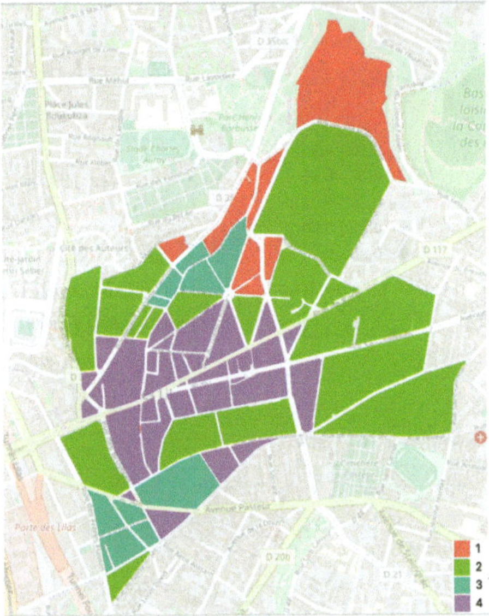

Fig. 2. City of Les Lilas after applying SOM algorithm at block level (3×3 SOM map and 4 instead clusters. We distinguish the suburban neighborhoods (super-cluster 1), the city center (super-cluster 4), and blocks composed of 60's building (super-cluster 2)

We can represent the nine prototypes in Fig. 3, which are 13-dimensional vectors. We can see that prototypes of opposite units 1 and 9 on SOM map (Fig. 3) are really different. Indeed, variable 2 "proportion of houses" has a low level for the prototype of unit 1, whereas it is has a high one for prototype of unit 9. Similarly, level of variable 3 "proportion of one-person households" is high for unit 1, and low for unit 9. And so on for the following variables. If we go from unit 1 to unit 9 passing through unit 5, we can notice that the prototype of unit 5 (which is at the center of SOM map) is pretty flat and thus, is similar to all prototypes of the map.

We set $\sigma = 600$ m and γ equal to the median of the distances between prototypes, using cross-validation.

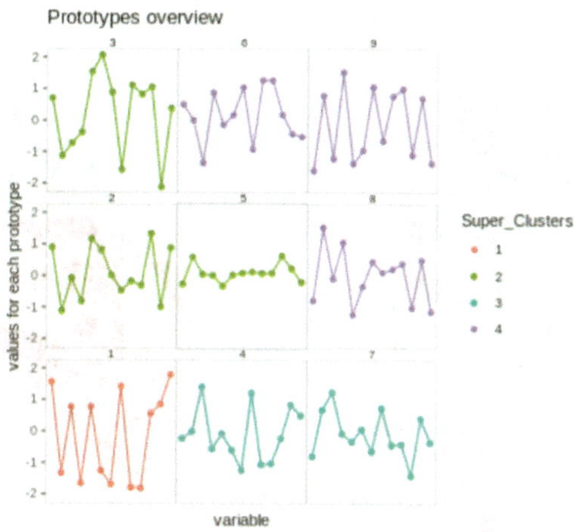

Fig. 3. Prototypes of the SOM map. In each unit, we represent a q-dimensional ($q = 13$) vector where components are: the proportion of individuals between 25 and 65 years old, proportion of houses, proportion of one-person households, proportion of 5 persons or more households, proportion of households below the low income threshold, proportion of households in collective housing, proportion of owner households, total number of individuals, total number of households, average winsorized tax income of individuals, average area of the primary residence (in m^2).

While a simple GWR only captures spatial effects and, with a delay, neighbourhood quality information reflected in the housing prices, our method captures the latter information before prices of actual transactions come to reflect it – allowing one to forecast future trends at a fine-grained geographical scale. Figure 4b is the combination of information unveiled by the Kohonen algorithm and a single GWR.

The differences (Fig. 4c) between price indices produced by pure GWR (Fig. 4a) versus our combined method (Fig. 4b) reflect information that is hardly accessible by other means, especially if one does not or cannot have an intimate knowledge of the city under consideration: the type and quality of the buildings, the *atmosphere* of a neighbourhood, whether it will soon be a very sought-after neighbourhood or not, *etc.*. Vast amounts of socio-demographical data work as a proxy for such information, provided one is able to harness it using machine learning methods.

6 Performance

We make the assumption that the real estate market does not have a significant fluctuation within 6 months and thus take into account transactions between 2019-09-01 and 2020-02-01 for the test set ($n = 70$ observations).

Performance of the model is measured by computing the relative error with respect to prices of observed transactions in the test set. We compute the pre-

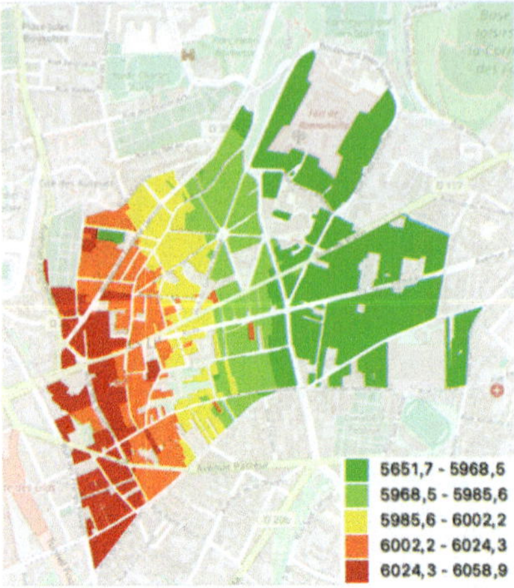

(a) Price map of Les Lilas obtained with a simple GWR.

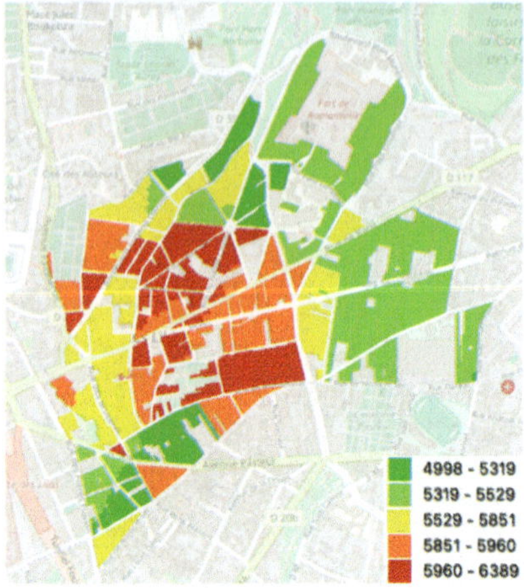

(b) Price map of Les Lilas obtained with the new model.

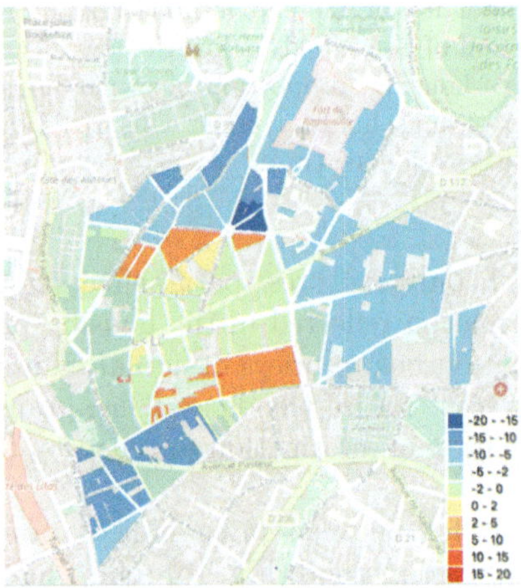

(c) Difference in percentage between a and b.

Fig. 4. Price maps.

diction error for both the GWR and the new model. At time t, $P_i(t)$ is the price of transaction i. We define the relative error such as

$$E = \frac{|\hat{P_i}(t) - P_i(t)|}{P_i(t)}, \tag{10}$$

where $\hat{P_i}(t)$ is the predicted price (Fig. 5).

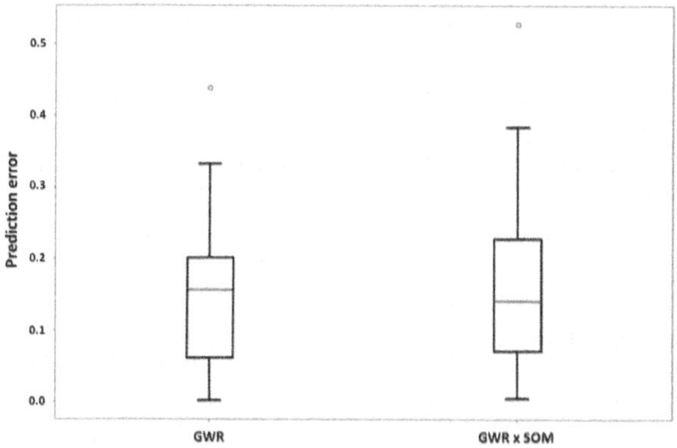

Fig. 5. Boxplots of prediction error of a GWR and of the new model measured on the test set ($n = 70$)

7 Conclusion

Using SOM allows one to gather information from a vast corpus of socio-economical data in order to bring out the socio-spatial structure and relate it to the dynamics of real estate prices. Combining distances on the Kohonen map with geographical distances provides a better model of prices (at least in the real estate markets where we have tested our method). Our method captures a reality that is hard (or expensive) to obtain via other, human-resource based practices.

Open questions include that of the definition of mixing weights between the two types of distances used, and that of the interpretability of regression coefficients thus obtained.

In future work, we will seek to develop a graph method in order to learn the size of the neighborhoods present in a city.

References

1. Bitter, C., Mulligan, G.F., Dall'erba, S.: Incorporating spatial variation in housing attribute prices: a comparison of geographically weighted regression and the spatial expansion method. J. Geog. Syst. **9**, 7–27 (2007). https://doi.org/10.1007/s10109-006-0028-7
2. Boelaert, J., Bendhaiba, L., Olteanu, M., Villa-Vialaneix, N.: SOMbrero: an R package for numeric and non-numeric self-organizing maps. In: Villmann, T., Schleif, F.-M., Kaden, M., Lange, M. (eds.) Advances in Self-Organizing Maps and Learning Vector Quantization. AISC, vol. 295, pp. 219–228. Springer, Cham (2014). https://doi.org/10.1007/978-3-319-07695-9_21
3. Fotheringham, A.S., Brunsdon, C., Charlton, M.: Geographically Weighted Regression: The Analysis of Spatially Varying Relationships. Wiley, Hoboken (2002)
4. Goodman, T.: Housing market segmentation and hedonic prediction accuracy. J. Hous. Econ. **12**(3), 181–201 (2003). https://EconPapers.repec.org/RePEc:eee:jhouse:v:12:y:2003:i:3:p:181-201

5. Kohonen, T.: Self-organizing Maps. Springer Series in Information Sciences, Springer, Heidelberg (2012)
6. Charlton, M., Fotheringham, A., Brunsdon, C.: Geographically weighted regression. J. Roy. Stat. Soc. Ser. D (The Statistician) 5–6 (2009)
7. Olteanu, M., Hazan, A., Cottrell, M., Randon-Furling, J.: Multidimensional urban segregation: toward a neural network measure. Neural Comput. Appl. **32**(24), 18179–18191 (2019). https://doi.org/10.1007/s00521-019-04199-5
8. Rosen, S.: Hedonic prices and implicit markets: product differentiation in pure competition. J. Polit. Econ. **82**, 34–55 (1974)
9. Se Can, A., Megbolugbe, I.: Spatial dependence and house price index construction. J. Real Estate Finance Econ. **14**, 203–222 (1997). https://doi.org/10.1023/A:1007744706720
10. Thouvenin: La formation des prix des logements anciens, les apports de la théorie des prix hédoniques. Books on Demand (2010)

Modification of the Classification-by-Component Predictor Using Dempster-Shafer-Theory

Mehdad Mohannazadeh Bakhtiari(✉) and Thomas Villmann

Saxon Institute for Computational Intelligence and Machine Learning, University of Applied Sciences Mittweida, Mittweida, Germany
m.mohannazade@gmail.com

Abstract. In this paper we propose a modification of the Classification-by-Components (CbC) model for classification. Particularly, we integrate explicitly the Dempster-Shafer-theory, which in the original approach was mentioned to be implicitly realized but not explained deeply. Thus, we redefine the CbC keeping the main aspects of positive and negative reasoning about detected components/features and relate this to the Demspster-Shafer-theory of evidence.

1 Introduction

Classification-by-Component (CbC) [4] is powerful extension of prototype based Learning Vector Quantization (LVQ) methods [2,5]. In CbC, the prototypes no longer require to represent classes, but rather they provide evidences for likelihood of the occurrence of a class. If we look at prototypes as independent sources of evidence, then each prototype would give a local estimate of class probabilities. With this point of view, if we want to have a global predictor, then the local evidences (given by each prototype) should be combined using a combination rule. One such combination rule is Dempster's combination rule [7], that is part of a framework called Dempster-Shafer-Theory (DST) [6]. In original CbC, the predictor is inspired by Dempster's combination rule and the neutral reasoning parameter is supposed to take care of conflict cases. However, a more careful formulation of CbC predictor may help with the success of the method.

In this paper, we offer a reformulation of CbC, such that the evidences, coming from each prototype, are combined using Dempster's combination rule. Also, we are interested in redefining the reasoning parameters, so they represent more meaningful quantities. The loss function that is compliant with the new formulation is probabilistic contrastive loss function [3]. The contrastive loss would greatly help with the derivation of gradients.

The rest of this paper is written as follows. Original CbC is briefly introduced in Sect. 2. In the same section, we discuss concerns about CbC more accurately. Section 3 gives a concise introduction to DST, with the emphasis on the parts

J. Faigl et al. (Eds.): WSOM+ 2022, LNNS 533, pp. 41–52, 2022.
https://doi.org/10.1007/978-3-031-15444-7_5

that are used in this paper. Section 4 would contain the main contribution of this paper, where CbC is reformulated, based on DST, to take care of the raised concerns. In Sect. 4, we have included the results of conducted simulations. Section 5 would conclude the paper.

2 Original Classification-by-Component

We are given a training set $T = \{(\mathbb{x}_i, y_i)\}_{i=1}^{\tau}$, such that $\mathbb{x}_i \in \mathbb{R}^m$ and $y_i \in S = \{1, 2, \ldots, c\}$. Also, there are prototypes $W = \{\mathbb{w}_t\}_{t=1}^{n}, \mathbb{w}_t \in \mathbb{R}^m$ in the feature space. The prototypes are called components as well. Detection probability function is defined, for a prototype $\mathbb{w}_t$, as a differentiable function $d_t : \mathbb{x} \to [0, 1]$. If $d_t(\mathbb{x}) = 1$, then component $\mathbb{w}_t$ has been fully detected in data $\mathbb{x}$. On the other hand, if $d_t(\mathbb{x}) = 0$, then component $\mathbb{w}_t$ is fully absent in data $\mathbb{x}$. The possibilistic class predictor of CbC, for class j, is

$$q_j(\mathbb{x}) = \frac{\sum_t r_{tj}^+ \cdot d_t(\mathbb{x}) + r_{tj}^- \cdot (1 - d_t(\mathbb{x}))}{\sum_t 1 - r_{tj}^0} \tag{1}$$

Here, the reasoning parameters are defined. r_{tj}^+ is the probability that component $\mathbb{w}_t$ is important for class j, i.e., it should be detected to support occurrence of j. r_{tj}^- is the probability that component $\mathbb{w}_t$ is important for class j but it should not be detected to support the occurrence of j. r_{tj}^0 is the probability that the component $\mathbb{w}_t$ is not important for class j.

The loss function of CbC is the contrastive loss, below, that is based on a monotonically increasing function Φ

$$l(\mathbb{x}, y) = \Phi(max\{q_i(\mathbb{x}) | i \neq y, i \in S)\} - q_y(\mathbb{x})) \,. \tag{2}$$

Despite the success of the method, based on convolutional neural network feature extractors, there are two main issues with the original formulation of CbC. Firstly, it is hard to show that the reasoning parameters of CbC converge to the their defined quantity. Secondly, although CbC predictor (1) is inspired by DST, the act of combining the evidences, coming from components, has not been carried out carefully.

3 Dempster-Shafer-Theory and Dempster's Combination Law

In real world machine learning problems, it is often the case that we cannot confidently form a traditional probabilistic space, mainly due to these three facts: lack of information, the fact that forming a complete space is not computationally efficient, and there are more than one source of information. Learning vector Quantization (LVQ) methods [2,5] usually deploy prototypes in the feature space to measure quantities and make decisions based on the information. One prototype, on its own, cannot capture the complexity of the space and, therefore, we need more than one prototypes. Each prototype is a source of information and information from each prototype should be meaningfully combined for a unified decision in the end. Traditional probability theory is not a suitable candidate for such scenarios. A solution is the Dempster-Shafer-Theory (DST), which is a generalization of traditional probability theory and it has been developed to measure uncertainty in scenarios, where knowledge is not complete. Knowledge is complete when we can form a joint distribution that describes the randomness in the process. The reader is referred to [6] for an introduction to DST concepts.

The core of DST is a mass function [1,6] $m : P(\phi) \rightarrow [0,1]$, that is defined on the power set $P(\phi)$ of the set of possible outcomes ϕ with

$$m(\emptyset) = 0 \,, \tag{3}$$

$$\sum_{A \in P(\phi)} m(A) = 1 \,, \tag{4}$$

where $\emptyset$ is the empty set. Note that, $A \subset B$, such that $A, B \in P(\phi)$, does not imply any information about the masses $m(A)$ and $m(B)$. Hence, $m(B)$ may be smaller, greater, equal to $m(A)$.

Based on the defined masses, functions such as belief and plausibility of events can be calculated. However, we would not focus on them, since they are of no interest in this paper. We are interested in Dempster's combination law that uses masses from two different sources and unify them to have a single mass function. Assume a set ϕ and its corresponding power set $P(\phi)$. Also, we assume two mass functions $m_1(.)$ and $m_2(.)$ defined on the power set $P(\phi)$, that satisfy (3) and (4). Dempster's combination law gives the combined mass

$$m_{\{1,2\}}(A) = \frac{\sum_{B \cap C = A \neq \emptyset} m_1(B) \cdot m_2(C)}{1 - \sum_{B \cap C = \emptyset} m_1(B) \cdot m_2(C)} \tag{5}$$

defined on the power set $P(\phi)$. The term $\sum_{B \cap C = \emptyset} m_1(B) \cdot m_2(C)$, appearing in the above equation, is a measure of the conflict between the information of the two sources.

4 A Modified CbC Based on Dempster-Shafer Theory

In this section, a modification to CbC networks is offered. We assume a training set and prototypes exactly as defined in the previous section. Assuming a threshold $\theta \in [0,1]$, a component $\mathbb{w}_t$ can divide the feature space in two parts. More specifically, $D_t(\theta) = \{\mathbb{x} | d_t(\mathbb{x}) \geq \theta\}$ is the area in which $\mathbb{w}_t$ is detected and $N_t(\theta) = \{\mathbb{x} | d_t(\mathbb{x}) < \theta\}$ is the area in which $\mathbb{w}_t$ is absent.

For a component $\mathbb{w}_t$, we define the following local probabilistic predictor for class j as

$$p_t(j|\mathbb{x}) = p(j|D_t(\theta)) \cdot p(\mathbb{x} \in D_t(\theta)) + p(j|N_t(\theta)) \cdot p(\mathbb{x} \in N_t(\theta)) \tag{6}$$

using the law of total probability and the fact that $D_t(\theta) \cup N_t(\theta)$ gives the entire feature space.

We define

$$r_{tj}^{+} = p(j|D_t(\theta)) \tag{7}$$

being the conditional model probability that the class is j given the data belongs to $D_t(\theta)$ and

$$r_{tj}^{-} = p(j|N_t(\theta)) \tag{8}$$

is the model probability for the class is j given the data belongs to $N_t(\theta)$. Hence, unlike original CbC, we would not require r_{tj}^{+} and r_{tj}^{-} to sum up to one, since they indicate quantities that may be independent.

Although the number of classes are often more than two, we consider a one-versus-all classification model with respect to each class $j \in S$, for simplicity of calculations, when combining evidences. Hence,

$$p_t(j|\mathbb{x}) + p_t(\neg j|\mathbb{x}) = 1 \tag{9}$$

holds true for each class $j \in S$.

We usually have more than one prototype, that would amount to more than one local predictor (6). The local predictors would be combined, using Dempster's combination law (5), to have a global predictor. We start with two prototypes $\mathbb{w}_t$ and $\mathbb{w}_k$ first. According to Dempster-Shafer Theory (DST), we define the following masses, with respect to prototype $\mathbb{w}_t$ and the power set $P(\phi_j) = \{\emptyset, \{j\}, \{\neg j\}, \phi_j\}$ of set $\phi_j = \{j, \neg j\}$:

$$m_t(j|\mathbb{x}) = p_t(j|\mathbb{x}) \ , \tag{10}$$

$$m_t(\neg j|\mathbb{x}) = p_t(\neg j|\mathbb{x}) = 1 - p_t(j|\mathbb{x}) \ . \tag{11}$$

According to (9), (10) and (11) these masses sum up to one. Therefore, considering (4) and keeping in mind that $m_t(\emptyset|\mathbb{x}) = 0$ from (3), we obtain

$$m_t(\phi_j|\mathbb{x}) = 0 \tag{12}$$

to be valid. The masses m_t and m_k of prototypes $\mathbb{w}_t$ and $\mathbb{w}_k$ are to be combined, according to (5) which yields

$$m_{\{t,k\}}(j|\mathbb{x}) = \frac{m_t(j|\mathbb{x}) \cdot m_k(j|\mathbb{x})}{m_t(j|\mathbb{x}) \cdot m_k(j|\mathbb{x}) + m_t(\neg j|\mathbb{x}) \cdot m_k(\neg j|\mathbb{x})} \tag{13}$$

for class j, with respect to power set $P(\phi_j)$ and

$$m_{\{t,k\}}(\neg j|\mathbb{x}) = \frac{m_t(\neg j|\mathbb{x}) \cdot m_k(\neg j|\mathbb{x})}{m_t(j|\mathbb{x}) \cdot m_k(j|\mathbb{x}) + m_t(\neg j|\mathbb{x}) \cdot m_k(\neg j|\mathbb{x})} \tag{14}$$

for class k together with

$$m_{\{t,k\}}(\phi_j|\mathbb{x}) = 0 \tag{15}$$

vanishing. Note that in case we did not assume a one-versus-all classification (meaning the set was $S = \{1, 2, \ldots, c\}$ as opposed to $\phi_j = \{j, \neg j\}$), we would have had to consider many more terms in the above masses, because the power set $P(S)$ would have had more members than $P(\phi_j) = \{\emptyset, j, \neg j, \phi_j\}$. One can easily prove that the masses can be extended for the combination of more than two prototypes which yields

$$m_t(j|\mathbb{x}) = \frac{\prod_t m_t(j|\mathbb{x})}{\prod_t m_t(j|\mathbb{x}) + \prod_t m_t(\neg j|\mathbb{x})} \tag{16}$$

$$m_t(\neg j|\mathbb{x}) = \frac{\prod_t m_t(\neg j|\mathbb{x})}{\prod_t m_t(j|\mathbb{x}) + \prod_t m_t(\neg j|\mathbb{x})} \tag{17}$$

$$m_t(\phi_j|\mathbb{x}) = 0 \tag{18}$$

accordingly. Similar, if the masses $m_t(j|\mathbb{x})$ are from prototypes $\mathbb{w}_t$, where $t \in \{1, 2, \ldots, k\}$, then $\boldsymbol{t} = [1, 2, \ldots, k]^T$ is a vector that collects the indices t of all contributing prototypes $\mathbb{w}_t, t \in \{1, 2, \ldots, k\}$. We define the global predictor for a class j

$$p(j|\mathbb{x}) = m_{\boldsymbol{t}}(j|\mathbb{x}) \tag{19}$$

by means of (16).

If $(\mathbb{x}, y) \in T$ is a training pair, the loss function, to be optimized by stochastic gradient descent learning, is based on the probabilistic Kullback-Leibler divergence

$$L(\mathbb{x}, y) = z_y \cdot \log \frac{z_y}{p(y|\mathbb{x})} + z_{\neg y} \cdot \log \frac{z_{\neg y}}{p(\neg y|\mathbb{x})} - z_y \cdot \log \frac{z_y}{p(\neg y|\mathbb{x})} - z_{\neg y} \cdot \log \frac{z_{\neg y}}{p(y|\mathbb{x})} \tag{20}$$

where $[z_y, z_{\neg y}]^T = [1, 0]^T$ is defined to be the probabilistic label vector of data point $\mathbb{x}((\mathbb{x}, y) \in T))$, that is based on the crisp label y [3]. The above equation can be simplified to

$$L(\mathbb{x}, y) = -\log p(y|\mathbb{x}) + \log p(\neg y|\mathbb{x}) = \log \frac{p(\neg y|\mathbb{x})}{p(y|\mathbb{x})} , \tag{21}$$

which serves as the local loss for a stochastic gradient descent training for a given pair $(\mathbb{x}, y)$. Note that, in the local loss function (21), the right hand side can be written as

$$\log \frac{p(\neg y|\mathbb{x})}{p(y|\mathbb{x})} = \log \frac{1 - p(y|\mathbb{x})}{p(y|\mathbb{x})} = \log(\frac{1}{p(y|\mathbb{x})} - 1) , \tag{22}$$

and that we can take the inverse of (19) by means of

$$\frac{1}{p(j|\mathbb{x})} = 1 + \prod_t \frac{m_t(\neg j|\mathbb{x})}{m_t(j|\mathbb{x})} . \tag{23}$$

Finally, (10), (22), and (23) are used in the loss (21) to rewrite the local loss as

$$L(\mathbb{x}, y) = \sum_t \log p_t(\neg y|\mathbb{x}) - \sum_t \log p_t(y|\mathbb{x}) . \tag{24}$$

The gradient of loss (24) with respect to $\mathbf{w}_t$ is

$$\frac{\partial L}{\partial \mathbf{w}_t} = \frac{\partial L}{\partial p_t(y|\mathbf{x})} \cdot \frac{\partial p_t(y|\mathbf{x})}{\partial d_t(\mathbf{x})} \cdot \frac{\partial d_t(\mathbf{x})}{\partial \mathbf{w}_t}, \tag{25}$$

where

$$\frac{\partial L}{\partial p_t(y|x)} = -(\frac{1}{p_t(\neg y|\mathbf{x}) \cdot p_t(y|\mathbf{x})}). \tag{26}$$

The gradients of the loss (24), with respect to the reasoning parameters r_{tj}^+ and r_{tj}^-, are

$$\frac{\partial L}{\partial r_{tj}^+} = \frac{\partial L}{\partial p_t(y|\mathbf{x})} \cdot \frac{\partial p_t(y|\mathbf{x})}{\partial r_{tj}^+} \tag{27}$$

and

$$\frac{\partial L}{\partial r_{tj}^-} = \frac{\partial L}{\partial p_t(y|\mathbf{x})} \cdot \frac{\partial p_t(y|\mathbf{x})}{\partial r_{tj}^-}, \tag{28}$$

respectively, where $\frac{\partial L}{\partial p_t(y|\mathbf{x})}$ is defined in (26). Also, $\delta(j,i)$ is the Kronecker delta function. Note that we need to normalize r_{tj}^+ and r_{tj}^- every now and then, such that $\sum_j r_{tj}^+ = 1$ and $\sum_j r_{tj}^- = 1$ are satisfied. Also, note that the adaptable parameters consist of all prototypes $\mathbf{w}_t$ as well as the reasoning parameters r_{tj}^+ and r_{tj}^-.

5 A Realization of the Modified CbC

In this section, the theoretically defined CbC network is practically realized. We first start with determination of the parameters for the local probabilistic predictor $p_t(j|\mathbf{x})$ in (6). Given

$$Sg_\gamma(x) = \frac{1}{1+\exp(-\gamma \cdot x)} \tag{29}$$

as the sigmoid function, we assume

$$(\mathbf{x} \in D_t(\theta)) = Sg_\gamma(d_t(\mathbf{x}) - \theta) \tag{30}$$

such that

$$p(\mathbf{x} \in N_t(\theta)) = Sg_\gamma(\theta - d_t(\mathbf{x})) \tag{31}$$

is valid. Hence, (6) becomes

$$p_t(j|\mathbf{x}) = r_{tj}^+ \cdot Sg_\gamma(d_t(\mathbf{x}) - \theta) + r_{tj}^- \cdot Sg_\gamma(\theta - d_t(\mathbf{x})). \tag{32}$$

Based on this realization, the second gradient in the right hand side of (25) is

$$\frac{\partial p_t(y|\mathbf{x})}{\partial d_t} = \gamma \cdot (r_{ty}^+ - r_{ty}^-) \cdot Sg_\gamma(d_t(\mathbf{x}) - \theta) \cdot Sg_\gamma(\theta - d_t(\mathbf{x})) \tag{33}$$

and, hence, we have

$$\frac{\partial L}{\partial p_t(y|\mathbb{x})} \cdot \frac{\partial p_t(y|\mathbb{x})}{\partial d_t(\mathbb{x})} = -(\frac{\gamma}{p_t(\neg y|\mathbb{x}) \cdot p_t(y|\mathbb{x})}) \cdot (r_{ty}^+ - r_{ty}^-) \cdot Sg_\gamma(d_t(\mathbb{x}) - \theta) \cdot Sg_\gamma(\theta - d_t(\mathbb{x})) \tag{34}$$

as the product of the first two gradients in (25).

Further, based on the realization (32), the second gradients of right hand sides of (44) and (45) become

$$\frac{\partial p_t(y|\mathbb{x})}{\partial r_{tj}^+} = Sg_\gamma(d_t(\mathbb{x}) - \theta) \tag{35}$$

and

$$\frac{\partial p_t(y|\mathbb{x})}{\partial r_{tj}^-} = Sg_\gamma(\theta - d_t(\mathbb{x})) \, . \tag{36}$$

5.1 Numerical Considerations for a Stable CbC

Numerical experiments indicated that the probabilities $p_t(y|\mathbb{x})$ and $p_t(\neg y|\mathbb{x})$, appearing in the denominator of gradient (26), may cause instabilities when they tend to extreme values 0 or 1. In this section, we investigate two solutions to reduce/overcome this problem.

The first solution is simply to keep explicitly both $p_t(j|\mathbb{x})$ and $p_t(\neg j|\mathbb{x})$ in a range $[\epsilon, 1-\epsilon]$, where $0 < \epsilon \ll 1$ is a small constant. Accordingly, given that $p_t(j|\mathbb{x})$ and $p_t(\neg j|\mathbb{x})$ are already in the range $[\epsilon, 1-\epsilon]$, we would update adaptive parameters in each step, based stochastic gradient descent, only if the new probabilities $p_t(j|\mathbb{x})$ and $p_t(\neg j|\mathbb{x})$ would remain in the range. Yet, this approach limits the CbC algorithm. Furthermore, it requires a more careful initialization of adaptive parameters, so they meet the condition.

The second solution, that is in essence similar to the first solution, is to define a transformation on the local predictor (32), such that the probabilities $p_t(j|\mathbb{x})$ and $p_t(\neg j|\mathbb{x})$ avoid extreme values 0 and 1. A suitable candidate is a squashing transformation

$$f_{a,b}(x) = \frac{1}{1 + \exp\left(-\frac{x-b}{a}\right)} \tag{37}$$

with $a, b \in \mathbb{R}$ and $a > 0$ such that it is monotonically increasing with the range being the interval $(0, 1)$. We would like to choose a and b, such that the argument of the exponential in (37) satisfies $(x - b) = 0$ at $x = 0.5$, i.e., $b = 0.5$. In other words, we are centering the sigmoid-like transformation at $x = 0.5$. Further, the parameter a determines the squashed range according to

$$f_{a,b=0.5}(x) \in (\epsilon(a), 1 - \epsilon(a)) \quad \text{with } \epsilon(a) = \frac{1}{1 + \exp\left(\frac{1}{2a}\right)}$$

avoiding the numerical instabilities. In our numerical experiment we set $a = \frac{1}{4}$, see Fig. 1. Further we remark that $f_{a,b=0.5}(p) + f_{a,b=0.5}(q) = 1$ is valid for $0 < p, q < 1$ provided that their sum is $p + q = 1$.

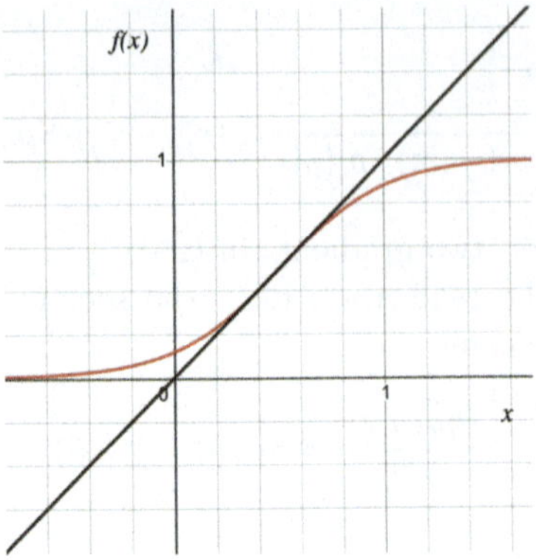

Fig. 1. Graph of transformation function $f_{\frac{1}{4},\frac{1}{2}}(x)$ in red. The identity function has been drawn in black for the sake of comparison.

By means of this squashing transformation we can approximate the local predictor (32) as

$$\hat{p}_t(j|\mathbf{x}) = f_{a,b=0.5}\left(p_t(j|\mathbf{x})\right) \tag{38}$$

satisfying the condition (9) and, consequently, we have

$$\hat{p}_t(\neg j|\mathbf{x}) = 1 - \hat{p}_t(j|\mathbf{x}) = f_{a,b=0.5}\left(p_t(\neg j|\mathbf{x})\right) \tag{39}$$

to be valid. Hence, we can re-write the contrastive loss function (24) as

$$\hat{L}(\mathbf{x}, y) = \sum_t \log \hat{p}_t(\neg y|\mathbf{x}) - \sum_t \log \hat{p}_t(y|\mathbf{x}) \tag{40}$$

with $p_t(y|\mathbf{x})$ being substituted by $\hat{p}_t(y|\mathbf{x})$ from (38). The gradient of $\hat{L}(\mathbf{x}, y)$ with respect to $\mathbf{w}_t$ is

$$\frac{\partial \hat{L}}{\partial \mathbf{w}_t} = \frac{\partial \hat{L}}{\partial \hat{p}_t(y|\mathbf{x})} \cdot \frac{\partial \hat{p}_t(y|\mathbf{x})}{\partial d_t(\mathbf{x})} \cdot \frac{\partial d_t(\mathbf{x})}{\partial \mathbf{w}_t}, \tag{41}$$

where

$$\frac{\partial \hat{L}}{\partial \hat{p}_t(y|\mathbf{x})} = -\left(\frac{1}{\hat{p}_t(\neg y|\mathbf{x}) \cdot \hat{p}_t(y|\mathbf{x})}\right) \tag{42}$$

and

$$\frac{\partial \hat{p}_t(y|\mathbf{x})}{\partial d_t} = 4 \cdot \hat{p}_t(y|\mathbf{x}) \cdot \hat{p}_t(\neg y|\mathbf{x}) \cdot \gamma \cdot (r_{ty}^{+} - r_{ty}^{-}) \cdot Sg_\gamma(d_t(\mathbf{x}) - \theta) \cdot Sg_\gamma(\theta - d_t(\mathbf{x})) \tag{43}$$

using $f_{a=\frac{1}{4}, b=\frac{1}{2}}$.

Note that, using transformation $f_{a,b}(x)$, the coefficient $\hat{p}_t(y|\mathbb{x}) \cdot \hat{p}_t(\neg y|\mathbb{x})$ in (43) cancels with $(\frac{1}{\hat{p}_t(\neg y|\mathbb{x}) \cdot \hat{p}_t(y|\mathbb{x})})$ in (42), when calculating the gradient (41). Hence, the gradients would not depend on the local probabilistic predictor $\hat{p}_t(j|\mathbb{x})$, that would otherwise cause numerical instabilities. Finally, the gradients of loss (40), with respect to the reasoning parameters r^+_{tj} and r^-_{tj}, are obtained as

$$\frac{\partial \hat{L}}{\partial r^+_{tj}} = \frac{\partial \hat{L}}{\partial \hat{p}_t(y|\mathbb{x})} \cdot \frac{\partial \hat{p}_t(y|\mathbb{x})}{\partial r^+_{tj}} \tag{44}$$

and

$$\frac{\partial \hat{L}}{\partial r^-_{tj}} = \frac{\partial \hat{L}}{\partial \hat{p}_t(y|\mathbb{x})} \cdot \frac{\partial \hat{p}_t(y|\mathbb{x})}{\partial r^-_{tj}} \tag{45}$$

with

$$\frac{\partial \hat{p}_t(y|\mathbb{x})}{\partial r^+_{tj}} = 4 \cdot \delta(j, y) \cdot \hat{p}_t(y|\mathbb{x}) \cdot \hat{p}_t(\neg y|\mathbb{x}) \cdot Sg_\gamma(d_t(\mathbb{x}) - \theta) \tag{46}$$

and

$$\frac{\partial \hat{p}_t(y|\mathbb{x})}{\partial r^-_{tj}} = 4 \cdot \delta(j, y) \cdot \hat{p}_t(y|\mathbb{x}) \cdot \hat{p}_t(\neg y|\mathbb{x}) \cdot Sg_\gamma(\theta - d_t(\mathbb{x})) \,. \tag{47}$$

6 Experimental Results

We applied the modified CbC network to MNIST data set, that has already been used to evaluate the performance of the original CbC [4]. MNIST data set consists of a training set of 60000 handwritten digits images and a test set of 10000 handwritten digits images, from 0 to 9. Labels, that are basically integers from 0 to 9, are provided for each image sample. We would normalize all the samples, such that the norm of each sample, as a vector, is equal to 1.

A convolutional Neural Network feature extractor, that considerably increases the classification accuracy, has been used in the original CbC. However, we would not use a feature extractor in our experiments.

As our first case, we apply the modified CbC, with 10 prototypes (components), to MNIST data set. The following initialization is done before the main training phase. We initialize 10 prototypes roughly at the center of samples with the same class. All prototypes would be normalized. Then, the prototypes are fixed and the reasoning parameters r^+_{tj} and r^-_{tj} would adapt to the current position of the prototypes for a couple of iterations. Then, we continue with the main training phase in a similar manner, where we fix either the prototypes or the reasoning parameters and train the other one for certain number of learning steps (100 learning steps). This alternating adaptation method allows the parameters r^+_{tj} and r^-_{tj}, that are statistics, to adapt based on more information (samples). After each learning step, the prototypes are normalized. This would keep the prototypes on the data manifold, as well as, it increases the performance drastically.

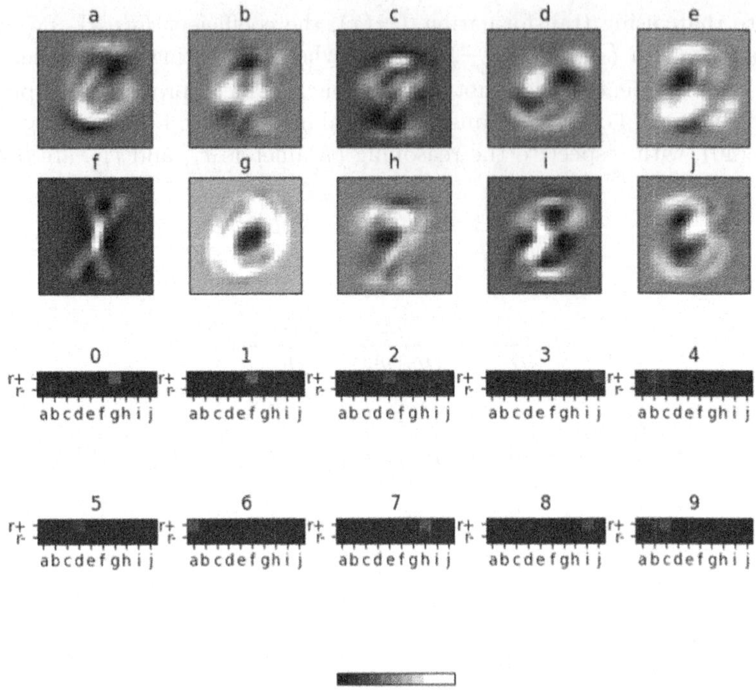

Fig. 2. Visualization of CbC with 10 components and the corresponding reasoning parameters. The first two rows show the learnt components a to j. The third and fourth rows show a gray scale representation of the reasoning matrices R_v, for a class v. The last row contains a gray scale map for 11 shades of gray from 0 to 1, with 0.1 space in between. The darkest shade represents value 0 and the lightest shade represents 1.

At this point, we may choose the detection function to be

$$d_t(\mathbb{x}) = \exp\left(-\frac{\|\mathbb{x} - \mathbb{w}_t\|^2}{2 \cdot \sigma^2}\right) \tag{48}$$

with the gradient

$$\frac{\partial d_t(\mathbb{x})}{\partial \mathbb{w}_t} = \frac{1}{\sigma^2} \cdot d_t(\mathbb{x}) \cdot (\mathbb{x} - \mathbb{w}_t) \tag{49}$$

and the parameter choice $\sigma = 1$.

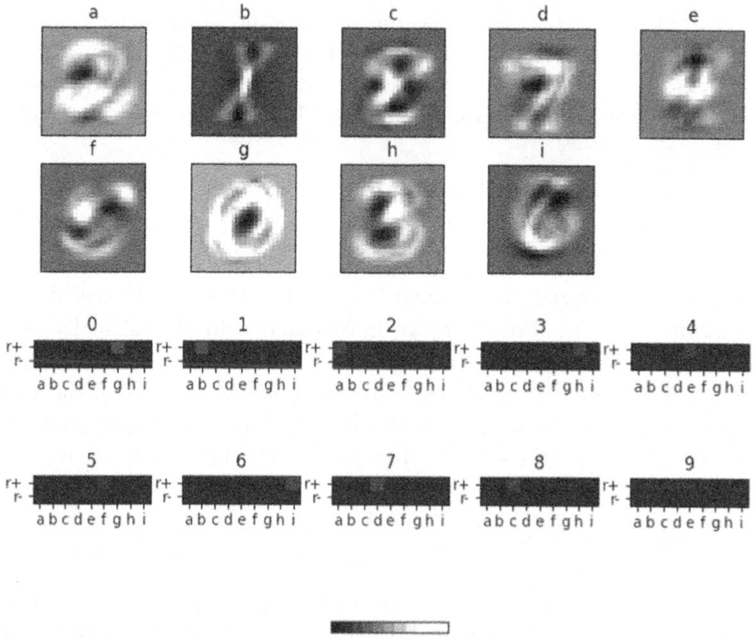

Fig. 3. Visualization of CbC with 9 components and the corresponding reasoning parameters. The first two rows show the learnt components a to i. The third and fourth rows show a gray scale representation of the reasoning matrices R_v, for a class v. The last row contains a gray scale map for 11 shades of gray from 0 to 1, with 0.1 space in between. The darkest shade represents value 0 and the lightest shade represents 1.

We also set the threshold $\theta = 0.9$ and the value of $\gamma = 1$ in (29) after grid search. The coefficient $Sg_\gamma(d_t(\mathbf{x}) - \theta) \cdot Sg_\gamma(\theta - d_t(\mathbf{x}))$, in (43), is nonzero only for data points $\mathbf{x}$ that roughly satisfy $d_t(\mathbf{x}) = \theta$, which are data points with detection level $\theta \pm \delta$. δ inversely depends on γ, such that a small γ would make a large δ and a large γ would decrease δ.

In order to avoid over-fitting, the training phase is stopped as soon as the classification accuracy starts to drop considerably. For MNIST data set with 10 prototypes, 10000 learning step achieve a high classification accuracy, that is $86.6\% \pm 0.2$, on test data, and hence less in original CbC but without convolutional layers here. In Fig. 2, we have visualized the components, as well as the reasoning parameters. Note that, for a class j, the reasoning parameters r^+_{tj} and r^-_{tj} are collected in a $2 \times k$ (k is the number of components) reasoning matrix R_j, such that the t-th column of the first row of R_j is r^+_{tj} and the t-th column of the second row of R_j is r^-_{tj}.

Thereafter we applied the modified CbC, with 9 components, to MNIST data set. The learning setting is as before, except for the following changes. Threshold parameter is set to $\theta = 0.95$, and the 9 components are initialized at the center of classes 0 to 8. The learning steps would automatically increase, as the number of parameters has reduced. In this case, 25000 learning steps would suffice to achieve a test data classification accuracy of 83.7% $\pm$ 0.1, which is again slightly behind original CbC with convolutional layers. The result of learning are visualized in Fig. 3.

7 Conclusion

In this paper, we have modified the CbC predictor, that is now based on Dempster's combination law. Moreover, we have made a mathematically consistent foundation for CbC networks, that is practically trainable and fast as well. The proposed modification has been tested by applying it to MNIST data set. The accuracy is similar to the original CbC. However, the developed mathematical foundation of this paper facilitates further investigations on CbC network.

References

1. Gordon, J., Shortliffe, E.H.: The Dempster-Shafer theory of evidence. Rule-Based Expert Syst. MYCIN Exp. Stanford Heuristic Program. Proj. **3**, 832–838 (1984)
2. Kohonen, T.: Learning vector quantization. In: Self-Organizing Maps. Springer Series in Information Sciences, vol. 30, pp. 175–189. Springer, Berlin, Heidelberg (1995). https://doi.org/10.1007/978-3-642-97610-0_6
3. Musavishavazi, S., Kaden, M., Villmann, T.: Possibilistic classification learning based on contrastive loss in learning vector quantizer networks. In: Rutkowski, L., Scherer, R., Korytkowski, M., Pedrycz, W., Tadeusiewicz, R., Zurada, J.M. (eds.) ICAISC 2021. LNCS (LNAI), vol. 12854, pp. 156–167. Springer, Cham (2021). https://doi.org/10.1007/978-3-030-87986-0_14
4. Saralajew, S., Holdijk, L., Rees, M., Asan, E., Villmann, T.: Classification-by-components: probabilistic modeling of reasoning over a set of components. In: Proceedings NeurIPS (2019)
5. Sato, A., Yamada, K.: Generalized learning vector quantization. In: NIPS, vol. 95, pp. 423–429 (1995)
6. Shafer, G.: A Mathematical Theory of Evidence. Princeton University Press (1976)
7. Shafer, G.: Dempster's rule of combination. Int. J. Approximate Reasoning **79**, 26–40 (2016)

Inferring ε-nets of Finite Sets in a RKHS

Antoine Moniot[(✉)], Isaure Chauvot de Beauchêne[(✉)], and Yann Guermeur[(✉)]

LORIA-CNRS, Campus Scientifique, BP 239,
54506 Vandœuvre-lès-Nancy Cedex, France
{Antoine.Moniot,Isaure.Chauvot-de-Beauchene,Yann.Guermeur}@loria.fr

Abstract. We introduce a method to derive ε-nets of finite sets. It operates in a reproducing kernel Hilbert space. Its principle combines two well-known tools of empirical inference: the hierarchical agglomerative clustering and the computation of minimum enclosing balls. It produces ε-nets whose cardinalities are smaller than those obtained with state-of-the-art methods.

Keywords: ε-nets · Empirical inference · RKHS · Hierarchical agglomerative clustering

1 Introduction

The inference of sets of prototypes is an ubiquitous problem in applied research. Such sets must satisfy two contradictory constraints: be representative and of cardinality as small as possible. When the data set is included in a Hilbert space (the criterion of fitness is a distance) and its cardinality is finite, then the problem reduces to that of the inference from a Gram matrix of an ϵ-net of minimum cardinality. This theoretical framework is of particular interest, for instance, in structural biology (use of rotamers [9]). Strangely enough, it still calls for the specification of dedicated methods. This is the subject of this study. Without loss of generality, it addresses the case when the dot product is provided by a kernel. Finding an optimal solution, i.e., an ϵ-net whose cardinality is the corresponding covering number, appears as a problem of combinatorial optimization which is intractable in practice. However, feasible solutions can be derived at negligible cost from the dendrograms produced by most of the major algorithms of hierarchical agglomerative clustering (HAC). In that context, a double observation can be made: the corresponding linkage functions are unevenly suited for the task, and it is possible to exhibit a new criterion a priori superior. Our method can be seen as a new HAC algorithm based on the computation of minimum enclosing balls (MEBs). When compared with state-of-the-art HAC algorithms on a classical benchmark, it produces uniformly the smallest ϵ-nets.

The organization of the paper is as follows. The problem is specified in Sect. 2. Section 3 is devoted to the state of the art. Our method is introduced in Sect. 4. Section 5 presents the experimental results. At last, we draw conclusions in Sect. 6.

J. Faigl et al. (Eds.): WSOM+ 2022, LNNS 533, pp. 53–62, 2022.
https://doi.org/10.1007/978-3-031-15444-7_6

2 Problem Characterization

The problem of interest is a problem of computational geometry. It is formalized as follows.

2.1 Theoretical Framework

We first characterize the data available. They constitute a finite subset of a description space $\mathcal{X}$. Let $s_{\mathcal{X}^n} = \{x_i : 1 \leqslant i \leqslant n\}$ denote this subset. A reproducing kernel Hilbert space (RKHS) [1] is spanned from $\mathcal{X}$ by means of a real-valued positive type function/kernel on $\mathcal{X}^2$. Let κ denote this function and $\left(\mathbf{H}_\kappa, \langle\cdot,\cdot\rangle_{\mathbf{H}_\kappa}\right)$ the corresponding RKHS. The canonical distance on $\mathbf{H}_\kappa$, ρ_κ, satisfies :

$$\forall (x, x') \in \mathcal{X}^2, \ \rho_\kappa (\kappa_x, \kappa_{x'}) = \|\kappa_x - \kappa_{x'}\|_{\mathbf{H}_\kappa} = \sqrt{\kappa (x, x) + \kappa (x', x') - 2\kappa (x, x')}.$$

This algebraic framework being established, the knowledge source available for inference can be defined as the Gram matrix $K \in \mathcal{M}_{n,n} (\mathbb{R})$ with entries $k_{i,j} = \kappa (x_i, x_j)$. The hypothesis of a Hilbert space structure and the availability of the matrix K are more than enough to tackle our problem of computational geometry (which becomes as a result a problem of functional analysis). This problem involves concepts whose introduction is usually attributed to Kolmogorov and Tihomirov [7]. Without loss of generality, they are presented below for metric spaces. $(\mathcal{E}, \rho)$ is such a space and $\mathcal{E}'$ is a subset of $\mathcal{E}$ which is totally bounded.

Definition 1 (ϵ-cover, ϵ-net and covering numbers).
For $\epsilon \in \mathbb{R}_+^*$, an ϵ-cover *of $\mathcal{E}'$ is a coverage of $\mathcal{E}'$ with open balls of radius ϵ the centres of which belong to $\mathcal{E}$. These centres form an* ϵ-net *of $\mathcal{E}'$. An* internal/proper ϵ-net *of $\mathcal{E}'$ is an ϵ-net of $\mathcal{E}'$ included in $\mathcal{E}'$. The* covering number *$\mathcal{N}(\epsilon, \mathcal{E}', \rho)$ is the smallest cardinality of its ϵ-nets:*

$$\mathcal{N}(\epsilon, \mathcal{E}', \rho) = \min \{|\mathcal{E}''| : (\mathcal{E}'' \subset \mathcal{E}) \wedge (\forall e \in \mathcal{E}', \ \rho(e, \mathcal{E}'') < \epsilon)\}.$$

$\mathcal{N}^{int}(\epsilon, \mathcal{E}', \rho)$ will designate a covering number of $\mathcal{E}'$ obtained by considering internal ϵ-nets only. We have thus:

$$\mathcal{N}^{int}(\epsilon, \mathcal{E}', \rho) = \min \{|\mathcal{E}''| : (\mathcal{E}'' \subset \mathcal{E}') \wedge (\forall e \in \mathcal{E}', \ \rho(e, \mathcal{E}'') < \epsilon)\}.$$

The problem considered is the following one.

Problem 1. Given the matrix K and $\epsilon \in \mathbb{R}_+^*$, find an ϵ-net $\mathcal{C}(\epsilon) = \{c_j(\epsilon) : 1 \leqslant j \leqslant |\mathcal{C}(\epsilon)|\}$ of $\mathcal{S} = \{\kappa_{x_i} : 1 \leqslant i \leqslant n\}$ with cardinality $|\mathcal{C}(\epsilon)|$ as close as possible to $\mathcal{N}(\epsilon, \mathcal{S}, \rho_\kappa)$.

2.2 Optimization Problem

Problem 1 can be turned into a problem of optimization by means of the following proposition, whose proof is straightforward.

Proposition 1. *The value of $\mathcal{N}(\epsilon, \mathcal{S}, \rho_\kappa)$ is $\mathcal{N}^*$ if and only if $\mathcal{N}^*$ is the minimum natural number such that there exists a partition of $\mathcal{S}$ into $\mathcal{N}^*$ subsets and for each subset, the radius of the MEB is inferior to ϵ.*

In the sequel, $\mathcal{P} = \{p_k : 1 \leqslant k \leqslant |\mathcal{P}|\}$ designates a partition of $\mathcal{S}$ into $|\mathcal{P}|$ subsests p_k. $\mathcal{P}(\mathcal{S})$ designates the set of all these partitions. For every $p_k \in \mathcal{P}$, $c(p_k)$ is the centre of its MEB. Then, $\mathcal{C}(\mathcal{P}) = \{c(p_k) : 1 \leqslant k \leqslant |\mathcal{P}|\}$. With this proposition and notation at hand, an optimal solution to Problem 1, i.e., a solution satisfying $|\mathcal{C}(\epsilon)| = \mathcal{N}(\epsilon, \mathcal{S}, \rho_\kappa)$, is obtained by solving Problem 2.

Problem 2.

$$\min_{\mathcal{P} \in \mathcal{P}(\mathcal{S})} |\mathcal{P}|$$

$$\text{s.t. } \forall k \in [\![1; |\mathcal{P}|]\!], \ \max_{\{i: \kappa_{x_i} \in p_k\}} \rho_\kappa(\kappa_{x_i}, c(p_k)) < \epsilon.$$

Let $\mathcal{P}^*$ denote an optimal solution to Problem 2. Then according to Proposition 1, $\mathcal{C}(\mathcal{P}^*)$ is an ϵ-net of $\mathcal{S}$ of minimal cardinality ($|\mathcal{P}^*| = \mathcal{N}(\epsilon, \mathcal{S}, \rho_\kappa)$), so that an optimal solution to Problem 1 is obtained by setting $\mathcal{C}(\epsilon) = \mathcal{C}(\mathcal{P}^*)$. Unfortunately, Problem 2 is an intractable problem of combinatorial optimization. To the best our knowledge, the only method available to solve it is the exhaustive way, and the number of partitions of $\mathcal{S}$, i.e., its Bell number B_n, is bounded from below by:

$$B_n \geqslant \left(\frac{n-1}{2}\right)^{\frac{n-1}{4}}.$$

However, the study of Problem 2 remains enlightening to derive a method producing (suboptimal) solutions to Problem 1. Indeed, there is a well-known family of classification methods initially producing as many clusters as there are points (trivial feasible solution to Problem 2) and iteratively merging clusters (producing partitions of smaller and smaller cardinality) according to a dissimilarity measure: the family of HAC algorithms [5]. If one can be satisfied with a suboptimal solution to Problem 2, then these algorithms turn the problem into a tractable one since the number of partitions considered is now upper bounded by n. The question then becomes: how close to $\mathcal{N}(\epsilon, \mathcal{S}, \rho_\kappa)$ is the cardinality of the smallest partition (closest to the root) associated with an ϵ-cover? Of course, the answer depends on the nature of the dissimilarity measure (the *linkage* d_H). The following section discusses the connection between the linkages of the HAC algorithms and the radii of the balls enclosing the sets constituting the partitions.

3 State of the Art

In this section, $\mathcal{S}_1$ and $\mathcal{S}_2$ stand for two subsets of $\mathcal{S}$ among the candidates to be merged at some iteration of an HAC algorithm. As such, they are specific subsets of $\mathcal{S}$. We have seen that even though the HAC algorithms are not designed to infer

ϵ-nets, they can all be used for that purpose. It suffices, for each of the nested partitions, to compute the corresponding balls using the matrix K. Furthermore, there exist HAC algorithms which exhibit the following property.

Property 1. The linkage function d_H increases or keeps unchanged each time two clusters are merged and there exists a function f_H such that, for every pair $(\mathcal{S}_1, \mathcal{S}_2)$, $r(\mathcal{S}_1 \bigcup \mathcal{S}_2)$, the radius of the smallest ball enclosing $\mathcal{S}_1 \bigcup \mathcal{S}_2$, is bounded from above by $f_H(d_H(\mathcal{S}_1, \mathcal{S}_2), |\mathcal{S}_1|, |\mathcal{S}_2|)$.

Property 1 is of central importance since it introduces an explicit connection between the partitions generated and ϵ-nets. Intuitively, the methods which have it are all the more promising to produce small ϵ-nets as $f_H(d_H(\mathcal{S}_1, \mathcal{S}_2), |\mathcal{S}_1|, |\mathcal{S}_2|)$ is a sharper upper bound on $r(\mathcal{S}_1 \bigcup \mathcal{S}_2)$. Practically, it makes it possible to produce a (non trivial) feasible solution to Problem 2 without computing any ball, i.e., without using K (after the dendrogram has been produced). This amounts to solving Problem 3.

Problem 3. Given an HAC algorithm, the dendrogram it produces for K and $\epsilon \in \mathbb{R}_+^*$, find among the partitions $\mathcal{P}$ for which it can be established that the set $\mathcal{C}(\mathcal{P})$ is an ϵ-net of $\mathcal{S}$, the partition closest to the root.

Consequently, when a function f_H is available, a trade-off is to be found between accuracy and computational complexity. Indeed, solving Problem 3 provides us with a (low-cost) initial solution to Problem 2 that can a priori be improved by exploiting K to compute the balls associated with partitions closer to the root. On the other hand, these balls are obtained as the solution of a quadratic programming problem (see Sect. 4.2).

These observations drive us into studying the main HAC algorithms [2] in the light of Property 1. They are characterized (by means of their linkage) in Table 1.

Table 1. Main HAC algorithms

Method	Linkage d_H
Average	$d_A(\mathcal{S}_1, \mathcal{S}_2) = \frac{1}{\lvert\mathcal{S}_1\rvert\lvert\mathcal{S}_2\rvert} \sum_{(\kappa_x, \kappa_{x'}) \in \mathcal{S}_1 \times \mathcal{S}_2} \rho_\kappa(\kappa_x, \kappa_{x'})$
Centroid	$d_{Ce}(\mathcal{S}_1, \mathcal{S}_2) = \rho_\kappa\left(\frac{1}{\lvert\mathcal{S}_1\rvert} \sum_{\kappa_x \in \mathcal{S}_1} \kappa_x, \frac{1}{\lvert\mathcal{S}_2\rvert} \sum_{\kappa_{x'} \in \mathcal{S}_2} \kappa_{x'}\right)$
Complete	$d_{Co}(\mathcal{S}_1, \mathcal{S}_2) = \max_{(\kappa_x, \kappa_{x'}) \in \mathcal{S}_1 \times \mathcal{S}_2} \rho_\kappa(\kappa_x, \kappa_{x'})$
Minimax	$d_{mM}(\mathcal{S}_1, \mathcal{S}_2) = \min_{\kappa_x \in \mathcal{S}_1 \bigcup \mathcal{S}_2} \max_{\kappa_{x'} \in \mathcal{S}_1 \bigcup \mathcal{S}_2} \rho_\kappa(\kappa_x, \kappa_{x'})$
Single	$d_S(\mathcal{S}_1, \mathcal{S}_2) = \min_{(\kappa_x, \kappa_{x'}) \in \mathcal{S}_1 \times \mathcal{S}_2} \rho_\kappa(\kappa_x, \kappa_{x'})$

We first note that one single function d_H does not satisfy the monotonicity condition of Property 1: d_{Ce} [8]. We thus restrict the study to the four other linkages. In the sequel, for each subset $\mathcal{S}'$ of $\mathcal{S}$, $\mathrm{diam}_\kappa(\mathcal{S}')$ designates its diameter with respect to the metric ρ_κ. The simplest function f_H to exhibit is obtained for the Minimax linkage. Indeed, since $\mathcal{S}_1 \bigcup \mathcal{S}_2 \subset \text{conv}\left(\mathcal{S}_1 \bigcup \mathcal{S}_2\right)$,

$$\begin{aligned}\forall \left(\mathcal{S}_1, \mathcal{S}_2\right),\ r\left(\mathcal{S}_1 \bigcup \mathcal{S}_2\right) &= \min_{\kappa_x \in \text{conv}\left(\mathcal{S}_1 \bigcup \mathcal{S}_2\right)} \max_{\kappa_{x'} \in \mathcal{S}_1 \bigcup \mathcal{S}_2} \rho_\kappa\left(\kappa_x, \kappa_{x'}\right) \\ &\leqslant d_{mM}\left(\mathcal{S}_1, \mathcal{S}_2\right)\end{aligned}$$

so that we can set:

$$\forall \left(\mathcal{S}_1, \mathcal{S}_2\right),\ f_{mM}\left(d_{mM}\left(\mathcal{S}_1, \mathcal{S}_2\right), \left|\mathcal{S}_1\right|, \left|\mathcal{S}_2\right|\right) = d_{mM}\left(\mathcal{S}_1, \mathcal{S}_2\right).$$

We now turn to the Complete linkage.

$$\begin{aligned}\forall \left(\mathcal{S}_1, \mathcal{S}_2\right),\ r\left(\mathcal{S}_1 \bigcup \mathcal{S}_2\right) &\leqslant \frac{1}{\sqrt{2}} \text{diam}_\kappa\left(\mathcal{S}_1 \bigcup \mathcal{S}_2\right) \qquad (1)\\ &= \frac{1}{\sqrt{2}} \max_{\left\{\kappa_x, \kappa_{x'}\right\} \in \mathcal{S}_1 \bigcup \mathcal{S}_2} \rho_\kappa\left(\kappa_x, \kappa_{x'}\right) \\ &= \frac{1}{\sqrt{2}} \max_{\left(\kappa_x, \kappa_{x'}\right) \in \mathcal{S}_1 \times \mathcal{S}_2} \rho_\kappa\left(\kappa_x, \kappa_{x'}\right) \qquad (2)\\ &= \frac{1}{\sqrt{2}} d_{Co}\left(\mathcal{S}_1, \mathcal{S}_2\right).\end{aligned}$$

Inequality (1) is an application of Jung's theorem. The transition to (2) (which would be false if $\mathcal{S}_1$ and $\mathcal{S}_2$ were arbitrary subsets of $\mathcal{S}$), springs from the "monotonicity" of d_{Co}. Consequently, we can set

$$\forall \left(\mathcal{S}_1, \mathcal{S}_2\right),\ f_{Co}\left(d_{Co}\left(\mathcal{S}_1, \mathcal{S}_2\right), \left|\mathcal{S}_1\right|, \left|\mathcal{S}_2\right|\right) = \frac{1}{\sqrt{2}} d_{Co}\left(\mathcal{S}_1, \mathcal{S}_2\right).$$

This bound is tight unless we can exploit some knowledge regarding the kernel, for instance the dimensionality of the RKHS. As for the Average linkage, since the corresponding sum can be bounded from below by the maximum, applying exactly the same reasoning as for the Complete linkage produces:

$$\forall \left(\mathcal{S}_1, \mathcal{S}_2\right),\ f_A\left(d_A\left(\mathcal{S}_1, \mathcal{S}_2\right), \left|\mathcal{S}_1\right|, \left|\mathcal{S}_2\right|\right) = \frac{\left|\mathcal{S}_1\right| \left|\mathcal{S}_2\right|}{\sqrt{2}} d_A\left(\mathcal{S}_1, \mathcal{S}_2\right).$$

Regarding the Single linkage, the triangle inequality provides us with

$$\forall \left(\mathcal{S}_1, \mathcal{S}_2\right),\ \text{diam}_\kappa\left(\mathcal{S}_1 \bigcup \mathcal{S}_2\right) \leqslant \text{diam}_\kappa\left(\mathcal{S}_1\right) + d_S\left(\mathcal{S}_1 \bigcup \mathcal{S}_2\right) + \text{diam}_\kappa\left(\mathcal{S}_2\right).$$

As a consequence, one can prove by induction that

$$\text{diam}_\kappa\left(\mathcal{S}_1 \bigcup \mathcal{S}_2\right) \leqslant \left(\left|\mathcal{S}_1 \bigcup \mathcal{S}_2\right| - 1\right) d_S\left(\mathcal{S}_1, \mathcal{S}_2\right).$$

Then, another application of Jung's theorem allows us to define the function f_S as:

$$\forall \left(\mathcal{S}_1, \mathcal{S}_2\right),\ f_S\left(d_S\left(\mathcal{S}_1, \mathcal{S}_2\right), \left|\mathcal{S}_1\right|, \left|\mathcal{S}_2\right|\right) = \frac{\left|\mathcal{S}_1 \bigcup \mathcal{S}_2\right| - 1}{\sqrt{2}} d_S\left(\mathcal{S}_1, \mathcal{S}_2\right).$$

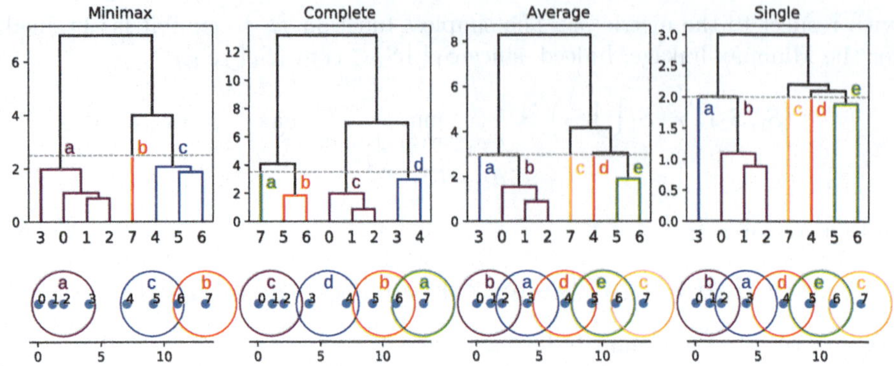

Fig. 1. Dendrograms produced with four linkages: d_{mM}, d_{Co}, d_A and d_S. A cut at the height of the dashed lines provides the corresponding solutions to Problem 3 (for $\epsilon = 2.5$). The resulting ϵ-covers are given below.

Figure 1 illustrates the solutions to Problem 3 obtained with the four linkage functions (d_{mM}, d_{Co}, d_A and d_S) on a set of 8 points of the real line.

The solution of smallest cardinality ($|\mathcal{P}| = 3$) is provided by the Minimax linkage. This solution is optimal since 3 is also the value of the covering number. Using K as an additional knowledge source to cut the dendrograms closer to the root produces new mergings so that eventually, all four ϵ-covers have minimal cardinality. Only one differs from the others: the one obtained with the Complete linkage.

Looking at the functions f_H, the Minimax linkage seems to be the best to produce ϵ-nets of small cardinality. It is noticeable that by definition, its ϵ-nets are internal, which represents an unnecessary restriction that could have a negative effect on the cardinality of the ϵ-net retained (keeping in mind that $\mathcal{N}(\epsilon, \mathcal{E}', \rho) \leqslant \mathcal{N}^{\text{int}}(\epsilon, \mathcal{E}', \rho)$). These observations are at the origin of the specification of our method, which is now introduced.

4 New Method of Inference

4.1 Definition

Our method for the empirical inference of ϵ-nets is basically the HAC algorithm involving the *radius* linkage function:

$$\forall (\mathcal{S}_1, \mathcal{S}_2),\ d_R(\mathcal{S}_1, \mathcal{S}_2) = r\left(\mathcal{S}_1 \bigcup \mathcal{S}_2\right). \tag{3}$$

The only difference is that the algorithm stops as soon as the candidate merging (minimizing d_R) produces a set whose MEB has a radius superior or equal to ϵ. The prototypes, i.e., the centres of the balls associated with the smallest partition reached, live in a RKHS which can be infinite dimensional (for instance if κ is a radial basis function). However, this raises no difficulties thanks to the kernel trick.

4.2 Derivation of the Prototypes

Since the prototypes are the centres of MEBs, each of them is obtained as the solution of an instance of a unique convex quadratic programming (QP) problem. The study of this problem provides us with their analytical expression (in the RKHS). Let $s_{\mathcal{X}^m}$ be a subset of $s_{\mathcal{X}^n}$ of cardinality m. Without loss of generality, its points are reindexed so that we can write $s_{\mathcal{X}^m} = \{x_i : 1 \leqslant i \leqslant m\}$. $K_m \in \mathcal{M}_{m,m}(\mathbb{R})$ is the corresponding Gram matrix and $\boldsymbol{\kappa}_m \in \mathbb{R}_+^m$ is its first diagonal. Then the QP problem is the following one:

Problem 4.

$$\min_{O \in \mathbf{H}_\kappa, R_2 \in \mathbb{R}_+} R_2$$

$$\text{s.t. } \forall i \in [\![1; m]\!], \ \rho_\kappa^2(O, \kappa_{x_i}) \leqslant R_2.$$

Problem 4 can be difficult to solve directly, especially if the RKHS is infinite dimensional. However, the classical way to infer the values of the parameters of a kernel method, the application of the Lagrangian duality, is available. The dual takes the following form:

Problem 5.

$$\min_{\alpha \in \mathbb{R}^m} \left\{ \alpha^T K_m \alpha - \boldsymbol{\kappa}_m^T \alpha \right\}$$

$$\text{s.t. } \begin{cases} \forall i \in [\![1; m]\!], \ \alpha_i \geqslant 0 \\ \mathbf{1}_m^T \alpha = 1 \end{cases}.$$

By noting (O^*, R_2^*) the solution to the primal problem and α^* the solution to its dual, we deduce from the Kuhn-Tucker (KT) conditions the analytical expression of the prototype of $s_{\mathcal{X}^m}$:

$$O^* = \sum_{i=1}^m \alpha_i^* \kappa_{x_i}. \tag{4}$$

Precisely, Eq. (4) provides the location of the centre of the MEB of $s_{\mathcal{X}^m}$ in its convex hull. With this formula at hand, the square distance between a prototype and any point in $\mathbf{H}_\kappa$ (let it be in $\mathcal{S}$ or not) is

$$\begin{aligned} \rho_\kappa^2(O^*, \kappa_x) &= \|O^*\|_{\mathbf{H}_\kappa}^2 + \|\kappa_x\|_{\mathbf{H}_\kappa}^2 - 2\langle O^*, \kappa_x \rangle_{\mathbf{H}_\kappa} \\ &= \sum_{i=1}^m \sum_{j=1}^m \alpha_i^* \alpha_j^* \kappa(x_i, x_j) + \kappa(x, x) - 2 \sum_{i=1}^m \alpha_i^* \kappa(x_i, x). \end{aligned}$$

As for the value of the radius $\sqrt{R_2^*}$, since $\mathbf{1}_m^T \alpha^* = 1$, the KT complementary conditions provide us with:

$$R_2^* = \sum_{i=1}^m \alpha_i^* \rho_\kappa^2(O^*, \kappa_{x_i}).$$

4.3 Algorithmic Implementation

The literature provides us with several exact algorithms to compute an MEB (see [6] for a survey). Their time complexity is a $O\left(m^3\right)$. However, they only apply to data sets living in a finite dimensional space (they solve Problem 4). This is precisely to bypass this limitation that we solve the dual (Problem 5) instead. It is the minimization of a quadratic form over the standard/probability simplex. This property can be exploited by a classical QP method, the Frank-Wolfe algorithm [4], which happens to be among the state-of-the-art "approximation algorithms" for the computation of MEBs. At each iteration, the solution of the linear programming problem is the vertex of the standard simplex whose index is the one of the smallest component of the gradient. Furthermore, there is an analytic expression for the optimal step size. Our implementation of the method makes use of both features. In practice, for a value of the duality gap equal to 0.99, the time complexity observed is a $O\left(m^2\right)$.

5 Experimental Results

Our method is evaluated in the framework of a comparative study. The criterion of assessment is the cardinality of the ϵ-nets generated. The methods of reference are the four HAC algorithms of Sect. 3 which exhibit Property 1. Thus, the study compares the efficacy (to produce small ϵ-nets) of the following linkage functions: d_R (given by Eq. 3), d_A, d_{Co}, d_{mM} and d_S (gathered in Table 1). Our method is implemented as described in Sect. 4. As for the algorithms from the literature, the ϵ-nets are obtained by solving Problem 3 based on the values returned by the corresponding functions f_H.

The benchmark selected is one of those used by Bien and Tibshirani in [2]: the Olivetti Faces data set[1]. It is made up of images of the faces of 40 different individuals. For each individual, 10 images are provided. They are 8-bit grayscale images of size 64×64. Thus, the data are vectors of $\mathbb{R}^{64 \cdot 64}$. This space is endowed with its canonical structure of Euclidean space. To sum up, the algorithms operate on the Gram matrix of $s_{\mathcal{X}^n} = \{x_i : 1 \leqslant i \leqslant n\}$ where $n = 400$. It is possible to infer from this matrix noteworthy elements: the two closest images, the two most distant images, the center of the MEB, and the images on its surface. Those elements are represented in Fig. 2.

Let $d_{2,\min} = \min_{1 \leqslant i < j \leqslant n} d_2(x_i, x_j)$. Then, $d_{2,\min}$, $r\left(s_{\mathcal{X}^n}\right)$, and $\operatorname{diam}\left(s_{\mathcal{X}^n}\right)$ can be used to restrict the study to the values of ϵ for which the derivation of an ϵ-net is non trivial. Indeed, if $\epsilon < \frac{1}{2} d_{2,\min}$, then $\mathcal{N}\left(\epsilon, s_{\mathcal{X}^n}, d_2\right) = n$ and we can set $\mathcal{C}\left(\epsilon\right) = s_{\mathcal{X}^n}$. Conversely, if $\epsilon > r\left(s_{\mathcal{X}^n}\right)$, then $\mathcal{N}\left(\epsilon, s_{\mathcal{X}^n}, d_2\right) = 1$ and we can set $\mathcal{C}\left(\epsilon\right) = c\left(s_{\mathcal{X}^n}\right)$ (centre of the MEB of $s_{\mathcal{X}^n}$). At last, if $\epsilon > \operatorname{diam}\left(s_{\mathcal{X}^n}\right)$, then $\mathcal{N}^{int}\left(\epsilon, s_{\mathcal{X}^n}, d_2\right) = 1$ and we can set $\mathcal{C}\left(\epsilon\right)$ equal to any singleton.

Consequently, the cardinalities of the ϵ-nets produced by the five algorithms are computed for ϵ in the interval $\left(0, \operatorname{diam}\left(s_{\mathcal{X}^n}\right)\right]$. The corresponding curves are gathered in Fig. 3.

[1] https://cs.nyu.edu/~roweis/data.html.

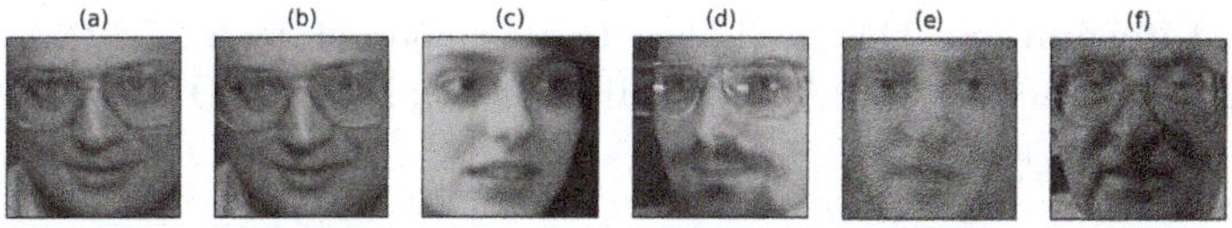

Fig. 2. Olivetti Faces dataset. (a) and (b): two closest images; (c) and (d): two most distant images; (e): center of the MEB; (f): one image on the surface of the MEB.

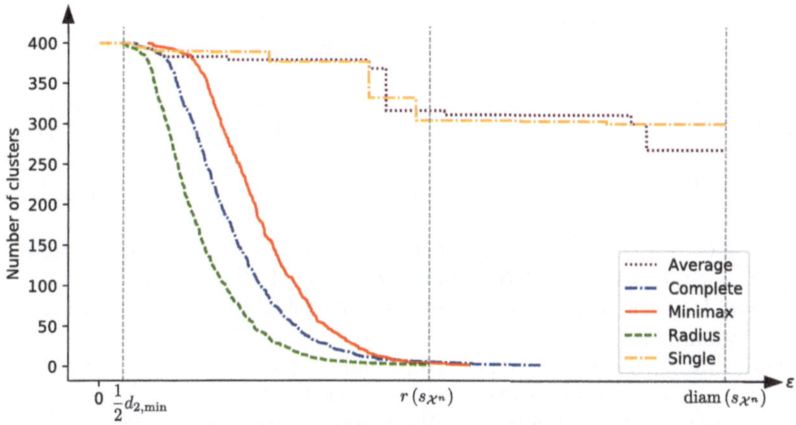

Fig. 3. Cardinality of the ϵ-nets produced by HAC algorithms as a function of ϵ

The new method is uniformly (for all values of ϵ) superior to the other four. Focusing on the HAC algorithms from the literature, it is noteworthy that contrary to our intuition, the Complete linkage appears to be more efficient than the Minimax linkage (at least for small values of ϵ). This phenomenon is the subject of ongoing investigations. As expected, neither d_A nor d_S appears as an appropriate choice.

6 Conclusions

A method to infer prototypes/ϵ-nets for finite subsets of a Hilbert space has been introduced. It can be seen as a hierarchical agglomerative clustering algorithm that operates in a reproducing kernel Hilbert space. Its linkage function takes in input two clusters of a partition and returns the radius of the minimum ball enclosing their union. The prototypes inferred can be easily handled in the RKHS, even when this space is infinite dimensional. A comparative experimental study involving the most challenging HAC algorithms from the literature suggests that our method produces smaller ϵ-nets, i.e., behaves as required.

Our current work is two-fold. It consists in performing additional comparative experiments and performing kernel engineering so as to dedicate the method to different applications in structural biology [3].

Funding Information. This work has been supported by the "RNAct" Marie Sklodowska-Curie Action (MSCA) Innovative Training Networks (ITN) H2020-MSCA-ITN-2018 (contract n° 813239).

References

1. Berlinet, A., Thomas-Agnan, C.: Reproducing Kernel Hilbert Spaces in Probability and Statistics. Kluwer Academic Publishers, Boston (2004)
2. Bien, J., Tibshirani, R.: Hierarchical clustering with prototypes via minimax linkage. J. Am. Stat. Assoc. **106**(495), 1075–1084 (2011)
3. Delannoy, A., Moniot, A., Guermeur, Y., Chauvot de Beauchêne, I.: Feature extraction for the clustering of small 3D structures: application to RNA fragments. In: JOBIM 2021 (2021)
4. Frank, M., Wolfe, P.: An algorithm for quadratic programming. Naval Res. Logistics Quart. **3**(1–2), 95–110 (1956)
5. Gower, J.C., Ross, G.J.S.: Minimum spanning trees and single linkage cluster analysis. J. R. Stat. Soc. Ser. C-Appl. Stat. **18**, 54–64 (1969)
6. Källberg, L.: Minimum enclosing balls and ellipsoids in general dimensions. Ph.D. thesis, Mälardalen University (2019)
7. Kolmogorov, A., Tihomirov, V.: ϵ-entropy and ϵ-capacity of sets in functional spaces. Am. Math. Soc. Translations Ser. **2**(17), 277–364 (1961)
8. Kopp, B.: Hierarchical classification III: average-linkage, median, centroid, WARD, flexible strategy. Biom. J. **20**, 703–711 (1978)
9. Murray, L.J.W., Arendall, W.B., Richardson, D.C., Richardson, J.S.: RNA backbone is rotameric. Proc. Natl. Acad. Sci. **100**(24), 13904–13909 (2003)

Steps Forward to Quantum Learning Vector Quantization for Classification Learning on a Theoretical Quantum Computer

Alexander Engelsberger(✉), Ronny Schubert, and Thomas Villmann

Saxon Institute for Computational Intelligence and Machine Learning, University of Applied Sciences Mittweida, Mittweida, Germany
engelsbe@hs-mittweida.de

Abstract. This paper introduces concepts for quantum computing based learning vector quantization and prototype learning models. These concepts relate to current quantum computing patterns and quantum hardware. For this purpose, we introduce a new computing pattern for prototype updates and possible measurement strategies in the quantum computing regime. Further, we consider numerical errors which are induced by the theoretical model and their impact on the learning process.

1 Introduction

The theoretical concept of quantum computers have been introduced by Richard Feynman 1982 [4]. After years of development in theoretical quantum computing, first real quantum computing systems nowadays start to work in practice [23]. Influenced by quantum hardware improvements, quantum computing is developing into different directions, one of them is the rapidly rising field of quantum machine learning (QML) [18]. QML covers both directions machine learning for enhanced quantum computing as well as quantum computing accelerating machine learning processes [3]. A major benefit for quantum computing is expected to come from data encoding strategies [22] and search strategies [7]. Both aspects also influence QML.

Several classical machine learning approaches maybe related to quantum computing including support vector machines, feedforward networks, recurrent and Hopfield networks as well as clustering algorithms [1,2,13,14]. Kernel methods in relation to QML are considered in [12,16,17]. Nearest neighbor approaches are studied in [8,24].

Generally, it is difficult to transfer complete machine learning algorithms to quantum devices or their digital twins. Currently, for many approaches rather

A.E. and R. S. are supported by grants of the European Social Fund and Agriculture Ministry of Germany.

J. Faigl et al. (Eds.): WSOM+ 2022, LNNS 533, pp. 63–73, 2022.
https://doi.org/10.1007/978-3-031-15444-7_7

the focus is to substitute time consuming conventional subroutines by quantum computing counterparts than to translate the full model [23].

Accordingly, first attempts were recently made to combine the learning vector quantization (LVQ) approach with quantum computing: LVQ aims to distribute prototype vectors in the data space for classification learning based on an attraction-repulsing-scheme (ARS) for prototype learning [9]. In [21], data as well as prototype vectors are consider as quantum states and learning is performed in the respective quantum state space. Yet, this approach runs on a classical computer whereas the math behind is inspired by quantum computing. In [20], a dissimilarity measure for quantum state vectors in context of LVQ was proposed that can be calculated on a quantum computer. However, the respective routine has to be run for each comparison between a prototype and a data vector slowing down the computational advantage of fast quantum calculations. Moreover, as discussed in [19], the repulsing scheme in LVQ constitutes a serious problem when transferring the LVQ-scheme to quantum computing.

The paper addresses both mentioned problems. We consider an extension of the so-called SWAP-test in quantum computing to calculate the dissimilarities between the quantum state input and all considered prototypes in parallel. Further, we give a possibility to realize the repulsing term in LVQ dynamics in quantum computing.

Although quantum computers have developed rapidly in the past years, only a few have just past the mark of 100 qubits. At the time of writing this paper, a computer system called Eagle by IBM[1] is the largest commercial quantum computer with 127 qubits. Generally available quantum computers and simulations for public use in research usually are significantly smaller and far away to be applicable to real problems in machine learning. Because of this limitations we restrict our self to theoretical considerations in this work.

2 Theoretical Framework

2.1 Learning Vector Quantization

Given a training data set $\mathcal{X} \subset \mathbb{R}^n$ with class labels $c(\boldsymbol{x}) \in \mathcal{C}$ for $\boldsymbol{x} \in \mathcal{X}$ and $|\mathcal{X}| = M_{\mathcal{X}}$, the goal of prototype learning for classification is to distribute a set of prototypes $\mathcal{W} = \{\boldsymbol{w}_1, \ldots, \boldsymbol{w}_{M_{\mathcal{W}}}\} \subset \mathbb{R}^n, c(\boldsymbol{w}_k) \in \mathcal{C}$, such that the classes are represented by the prototypes as best as possible such that the misclassification rate is minimized. The model prediction $c_{\mathcal{W}}(\boldsymbol{x})$ takes place as a nearest prototype decision according to a given dissimilarity measure $d(\boldsymbol{x}, \boldsymbol{w}_k)$, i.e., $c_{\mathcal{W}}(\boldsymbol{x}) = c(\boldsymbol{w}_{k^*})$ with $\boldsymbol{w}_{k^*}$ has minimum dissimilarity $d(\boldsymbol{x}, \boldsymbol{w}_{k^*})$ among all prototypes. Frequently, the dissimilarity is the squared Euclidean distance.

The family of learning vector quantization algorithms (LVQ) performs model adaptation for a given training sample $\boldsymbol{x} \in \mathcal{X}$ by an attraction repulsing scheme

[1] https://newsroom.ibm.com/2021-11-16-IBM-Unveils-Breakthrough-127-Qubit-Quantum-Processor.

[9]: A prototype $\boldsymbol{w}_k$ is moved towards $\boldsymbol{x}$ if the prototype class $c(\boldsymbol{w}_k)$ matches the data class $c(\boldsymbol{x})$. Otherwise, it is repelled. More specifically, we have

$$\Delta \boldsymbol{w}_k = -\Psi(\boldsymbol{x}) \cdot \epsilon \cdot \frac{\partial d(\boldsymbol{x}, \boldsymbol{w}_k)}{\partial \boldsymbol{w}_k} \tag{1}$$

with $0 < \epsilon \ll 1$ being the learning rate and $\Psi(\boldsymbol{x}) = 1$ for $c(\boldsymbol{w}_k) = c(\boldsymbol{x})$ and -1 elsewhere. For the squared Euclidean distance d_E^2 we obtain

$$\boldsymbol{w}_k \mapsto \boldsymbol{w}_k + \Psi(\boldsymbol{x}) \cdot \epsilon \cdot (\boldsymbol{x} - \boldsymbol{w}_k) = \Psi(\boldsymbol{x}) \cdot \epsilon \cdot \boldsymbol{x} + (1 - \Psi(\boldsymbol{x}) \cdot \epsilon) \cdot \boldsymbol{w}_k \tag{2}$$

as adaptation in case of class agreement, which can be seen as a linear interpolation (LERP) between $\boldsymbol{x}$ and $\boldsymbol{w}_k$ for $\Psi(\boldsymbol{x}) = 1$.

The simple but heuristic LVQ1 algorithm follows exactly this adaptation scheme for the best matching prototype $\boldsymbol{w}_{k^*}$ with

$$k^* = \operatorname{argmin}_{k=1}^{M_{\mathcal{W}}} (d(\boldsymbol{x}, \boldsymbol{w}_k)) \tag{3}$$

is denoted as winner-takes-all-rule (WTA-rule) [9]. More sophisticated LVQ-variants perform a stochastic gradient descent scheme for the approximated classification error [15].

In case of normalized data and prototypes with $\|\boldsymbol{x}\| = 1$ and $\|\boldsymbol{w}_k\| = 1$ the squared Euclidean distance becomes

$$d_E^2(\boldsymbol{x}, \boldsymbol{w}_k) = 2 - 2\,\langle \boldsymbol{x} | \boldsymbol{w}_k \rangle\,, \tag{4}$$

where the standard Euclidean inner product $\langle \boldsymbol{x} | \boldsymbol{w}_k \rangle$ can be interpreted as the excitation of a perceptron with the weight vector $\boldsymbol{w}_k$ given the stimulus $\boldsymbol{w}_k$. Thus the nearest prototype learning in LVQ is closely related to Hebbian learning. Further, we remark that the data and the prototypes belong to the unit hypersphere in this setting. For this configuration, angle-LVQ (ALVQ) was proposed in [5] using the dissimilarity

$$d_\alpha(\boldsymbol{x}, \boldsymbol{w}_k) = \arccos \langle \boldsymbol{x} | \boldsymbol{w}_k \rangle, \tag{5}$$

which yields the angle α between the vectors. Important to note, due to the normalization, $d_\alpha(\boldsymbol{x}, \boldsymbol{w}_k)$ is also equal to the arc length between the vectors. Accordingly, a corresponding prototype update can be defined as

$$\boldsymbol{w}_k \mapsto \frac{\sin(\Psi(\boldsymbol{x}) \cdot \epsilon \cdot \alpha)}{\sin(\alpha)} \cdot \boldsymbol{x} + \frac{\sin((1 - \Psi(\boldsymbol{x}) \cdot \epsilon) \cdot \alpha)}{\sin(\alpha)} \cdot \boldsymbol{w}_k, \tag{6}$$

as suggested in [5]. It can bee seen as a spherical linear interpolation (SLERP) for $\Psi(\boldsymbol{x}) = 1$, which is well-known in computer graphics for angle interpolation [11]. Note, if we have the value (angle) $\alpha = d_\alpha(\boldsymbol{x}, \boldsymbol{w}_k)$ before the update, after SLERP the value $\epsilon \cdot \alpha$ is valid for the angle.

2.2 How to Represent Classical Data on a Quantum Computer and Respective Quantum State Dissimilarity Calculation

Data encoding is a crucial aspect in quantum machine learning [10,22]. For our theoretical approach it is sufficient to define pure quantum states, a special subset of all states a quantum system can take. Note, quantum states are equivalently denoted as wave functions. A system of N qubits can be described by an *amplitude encoding*:

$$\psi = \sum_{i=0}^{2^N-1} x_i |i\rangle =: |\boldsymbol{x}\rangle \quad \text{with} \quad \sum_{i=0}^{2^N-1} |x_i|^2 = 1$$

for complex entries $x_i \in \mathbb{C}$. We use the notation $|i\rangle$ as an abbreviation that the quantum register is in the binary state representing the natural number i. The normalization condition is connected to probabilities: If $|\boldsymbol{x}\rangle$ is measured its probability of being in state i is $|x_i|^2$.

Multiple patterns for the encoding of classical data on a quantum computer are well known [18]. If data are given in a binary representation, M data points x_i can be represented in a superposition of their binary state as

$$\sum_{i=0}^{M} g_i |x_i\rangle \quad \text{with} \quad \sum_{i=0}^{M} |g_i|^2 = 1,$$

where g_i is a weight for each state. For this, so-called *basis encoding* strategy, the quantum register needs to contain as many qubits as the binary data representation has.

To represent our real valued input vectors we choose the amplitude encoding pattern. This encoding needs a small number of qubits to represent real numbers, by avoiding floating number encoding and by encoding N dimensions with $\log_2 N$ qubits. It also allows to apply routines like the later introduced SWAP-test. Any normalized vector can be transformed into a valid quantum state by padding it with 0 up to a power of two

$$[x_1, \ldots, x_n] \in \mathbb{R}^n \rightsquigarrow [x_1, \ldots, x_n, 0, \ldots, 0] \in \mathbb{R}^{2^N}$$

with $N = \lceil \log_2(n) \rceil$. An initialized quantum system with N qubits in this state now represents the vector. In the following, we assume that all vectors are normalized and padded accordingly. This encoding works for complex vectors and, hence, also for real $\mathcal{X} \subset \mathbb{R}^n$ and $\mathcal{W} \subset \mathbb{R}^n$ with $n = 2^N$.

According to the previous explanations we are able to transform data and prototype vectors into quantum states, which all are elements of the respective Bloch-sphere, i.e. they belong to the unit hypersphere.[2]

[2] An aspect to be considered for coding is that we need to distinguish states (vectors) by a respective measurement. For example, phase information cannot be identified: Suppose the system $a|0\rangle + b|1\rangle$ with amplitudes a, b. It cannot be separated from the system $ca|0\rangle + cb|1\rangle$ with a global phase c if the absolute value $|c| = 1$ is valid.

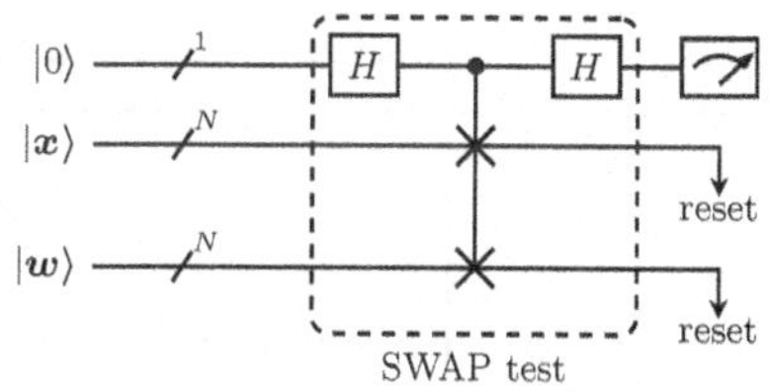

Fig. 1. SWAP-test to measure the dissimilarity between the states $|\mathbf{x}\rangle$ and $|\mathbf{w}\rangle$ using the ancilla qubit $|0\rangle$.

For nearest prototype learning like LVQ, the WTA-rule (3) requires to calculate all dissimilarities between a data point and the prototypes which are now all considered to represent quantum states. The squared Euclidean distance can be computed by means of the SWAP-test as an appropriate quantum computer routine: For this purpose, an input subsystem of size $2N$ is initialized into the state $|\boldsymbol{x}\rangle \otimes |\boldsymbol{w}\rangle$, where each vector is initialized independently on a quantum system of size N provided an efficient routine to prepare a state in amplitude encoding. The SWAP-test is applied to this system combined with a single ancilla qubit by the circuit shown in Fig. 1. For $N > 1$, a controlled SWAP-gate is applied to all N qubit pairs in both registers controlled by the same ancilla. The probability that the ancilla is in state $|0\rangle$ after the SWAP-test is

$$p_a(|0\rangle) = \frac{1}{2}\left(1 + |\langle \boldsymbol{x}|\boldsymbol{w}_k\rangle|^2\right)$$

and, hence, we can calculate

$$d_E^2(\boldsymbol{x}, \boldsymbol{w}_k) = 2 - 2\sqrt{2p_a(|0\rangle) - 1} \quad \text{and} \quad d_\alpha(\boldsymbol{x}, \boldsymbol{w}_k) = \arccos\sqrt{2p_a(|0\rangle) - 1}$$

as a quantum computing realizations of (4) and (5), respectively. Yet, this solution does not provide a parallel approach for multiple dissimilarity calculations as needed for an efficient LVQ winner determination.

3 Quantum Routines for LVQ

In this section, we provide the main results of the paper which are an extension of the usual SWAP-test such that an input can be compared with all prototypes in parallel and the realization of the attraction-repulsing scheme (ARS) for LVQ learning of the prototypes taken as quantum states.

3.1 Application of the SWAP-Test to Sets of Vectors

Recently, a quantum method for pairwise dissimilarity calculation was proposed in [6]. It is based on a pairing unitary operator that takes a set of input quantum registers and two indexing ancilla registers as input. The ancilla registers have a

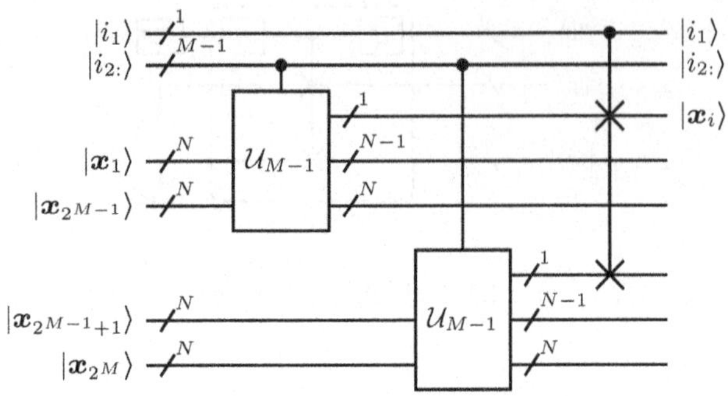

Fig. 2. Recursive construction of the selection unitary operator $\mathcal{U}_M$ for 2^M input registers $|x_i\rangle$.

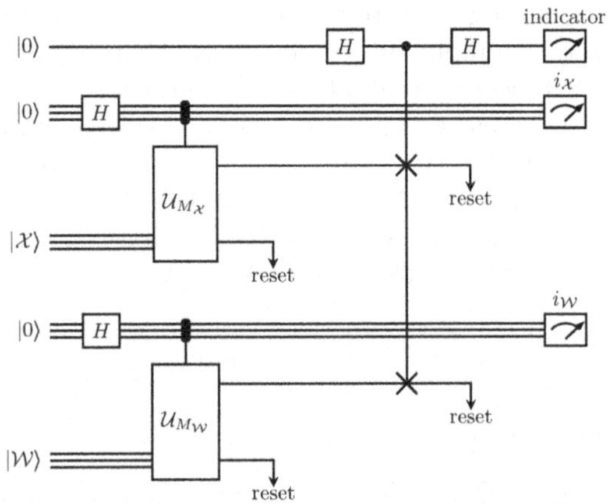

Fig. 3. The circuit of the SWAP-test for comparison of two state vector sets $|\mathcal{X}\rangle$ and $|\mathcal{W}\rangle$.

basis encoded index. The result of the application of this unitary is a permutation of the input registers such that the two indices are switched into the first two positions. Now, initializing both ancilla registers in a superposition of all possible indices and applying the pairing unitary operator we obtain the first two registers to be in a superposition of all pairs entangled with the corresponding indices in the ancilla registers. Thus, a SWAP-test between these two positions can be applied. The ancilla of the SWAP-test combined with the two index registers contains the inner product information of all pairs.

As already explained, the indexing of the prototypes and data are realized in the quantum variant by extra index registers. To adapt the previous pairing scheme to the multiple setting of the calculation of all dissimilarities between an input vector set and all prototypes required for the WTA-rule (3), we replace the single pairing unitary by the application of two selection unitaries, i.e. two operators each of them applied to a subsystem of input registers. Following this idea, a selection unitary for 2^M input registers can be recursively generated: After applying $\mathcal{U}_1$, which is a single controlled SWAP gate, the selected register has been swapped into the first position. Now we can extend this scheme by a recursive definition for $\mathcal{U}_M$ as depicted in Fig. 2 where i_1 is the most significant bit of i and $i_{2:}$ collects all other bits in binary representation. We combine two of these selection unitaries, one with $M = 2^{M_\mathcal{X}}$ for the data and one with $M = 2^{M_\mathcal{W}}$ for the prototypes, with the SWAP-test subroutine to obtain a circuit as given in Fig. 3 containing three ancilla registers. For this circuit, we consider the measurement probability of a state $|a\rangle\,|i_\mathcal{X}\rangle\,|i_\mathcal{W}\rangle$ for the three ancillas. The probability of measuring the ancilla registers in state $|0\rangle\,|j\rangle\,|k\rangle = |0jk\rangle$ is

$$p(\text{observing } |0jk\rangle) = \frac{1 + |\langle \boldsymbol{x}_j|\boldsymbol{w}_k\rangle|^2}{2^{N_\mathcal{X}+N_\mathcal{W}+1}},$$

where $N_\mathcal{X}$ and $N_\mathcal{W}$ are the size of the data and the prototype index registers, respectively. For a single stimulus $|\boldsymbol{x}\rangle$ we have $N_\mathcal{X} = 0$, therefore no index register is necessary and we get

$$p(\text{observing } |0k\rangle) = \frac{1 + |\langle \boldsymbol{x}|\boldsymbol{w}_k\rangle|^2}{2^{N_\mathcal{W}+1}}.$$

Additionally, for a single prototype we get $N_\mathcal{W} = 0$ as well and the probability reduces to the familiar SWAP-test

$$p(\text{observing } |0\rangle) = \frac{1 + |\langle \boldsymbol{x}|\boldsymbol{w}\rangle|^2}{2^1} = \frac{1}{2} + \frac{1}{2}|\langle \boldsymbol{x}|\boldsymbol{w}\rangle|^2.$$

3.2 qSLERP Action

We already established SLERP as a vector shift on a spherical manifold. Now, we introduce a quantum circuit that applies SLERP on two quantum states. The resulting quantum-SLERP (qSLERP) action applies an interpolation or extrapolation for the quantum states determined by the two amplitude vectors $\boldsymbol{x}, \boldsymbol{w}$ and a parameter $\epsilon \in \mathbb{R}$, i.e. it is an SLERP-operation in the quantum state space (Bloch-sphere).

To store one of these vectors in amplitude encoding mode, a quantum register of N qubits would be needed. Yet, to apply qSLERP, both vectors $\boldsymbol{x}$ and $\boldsymbol{w}$ are encoded in the same register by the use of an extra ancilla qubit for a total of $N + 1$ qubits

$$\psi_{\boldsymbol{x};\boldsymbol{w}} = \frac{1}{\sqrt{2}} \sum_{k=0}^{2^N-1} [x_k\,|0\rangle\,|k\rangle + w_k\,|1\rangle\,|k\rangle]. \tag{7}$$

The first qubit is the ancilla. The normalisation constant $\frac{1}{\sqrt{2}}$ is necessary to get a valid wave function.

Updating a prototype $\boldsymbol{w}$ on a quantum computer can be done by the circuit shown in Fig. 4 using the preparation $\psi_{\boldsymbol{x};\boldsymbol{w}}$. The circuit uses the update angle

$$\theta = 2\,\text{arctan2}(|\sin(\epsilon\alpha)|, |\sin((1-\epsilon)\alpha)|), \tag{8}$$

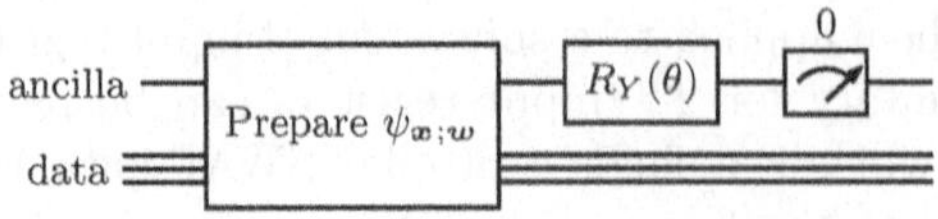

Fig. 4. qSLERP circuit based on an update angle θ.

where α is the current angle between the prototype $\boldsymbol{w}$ and data point $\boldsymbol{x}$. The SWAP-test can be used to calculate α. Note that we have to take care regarding the periodicity of the arctan2-function.

In the following we explain, how this simple qSLERP-circuit is working: When the data and prototype is loaded, the system takes the state vector

$$[x_0, \ldots, x_{2^N-1}, w_0, \ldots, w_{2^N-1}]^T.$$

Application of the unitary operator

$$U_{slerp} = \frac{1}{\zeta}\begin{bmatrix} \sin((1-\epsilon)\alpha) & \sin(\epsilon\alpha) \\ -\sin(\epsilon\alpha) & \sin((1-\epsilon)\alpha) \end{bmatrix}$$

to the ancilla, the state vector changes to

$$\frac{1}{\zeta}[\sin((1-\epsilon)\alpha)x_0 + \sin(\epsilon\alpha)w_0, \ldots, \sin((1-\epsilon)\alpha)x_0 - \sin(\epsilon\alpha)w_0, \ldots,]^T.$$

If we now apply a branch selection on the ancilla measured as 0, the state vector changes to

$$\frac{1}{\xi}[\sin((1-\epsilon)\alpha)x_0 + \sin(\epsilon\alpha)w_0, \ldots, \sin((1-\epsilon)\alpha)x_{2^N-1} + \sin(\epsilon\alpha)w_{2^N-1}, 0, \ldots]^T.$$

It can be shown that $\xi = \sin(\alpha)$, therefore the system represents the result of SLERP.

Whats left is the decomposition of U_{slerp} into elementary gates. Every unitary matrix can be decomposed into the three rotation gates RZ, RY, and RZ such that $U_{slerp} = R_z(\gamma_1) \cdot R_y(\theta) \cdot R_z(-\gamma_2)$. The inner RY is parametrized with θ as shown above paying attention to the periodicity of the arctan2-function. For real matrices U_{slerp} the values γ_1 and γ_2 for the outer RZ gates are dependent: For example i) $\gamma_1 = \gamma_2 = 0$, ii) $\gamma_1 = \pi$ and $\gamma_2 = -\gamma_1$, and iii) other configurations regarding the periodicity. Thus using this periodicity, the three gates can be replaced by a single RY gate as shown in Fig. 5. Here ii) refers to case 1 and iii) is represented by case 2. For i) the RZ -gates become the identity. In this sense, we can reduce the decomposition of U_{slerp} to a phase correction applied to the single RY gate.

4 Discussion

4.1 Numerical Errors Imposed by Approximation

For both SWAP-test and qSLERP, it is necessary to approximate the quantum state after the circuit execution. Strategies for this approximation belong to the

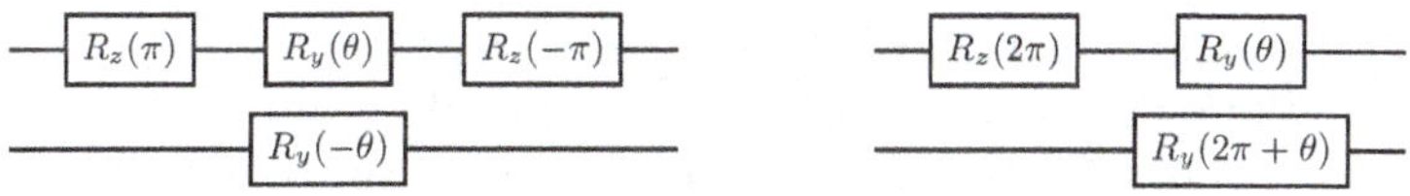

(a) Case 1 (b) Case 2

Fig. 5. ZYZ Decomposition: Gate replacements

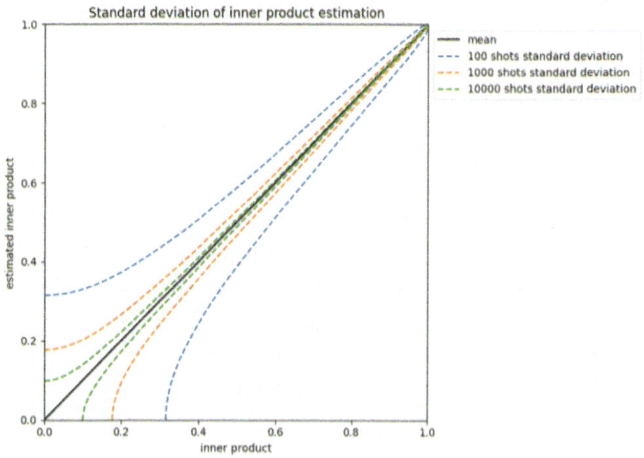

Fig. 6. Standard deviation based on the theoretical model of SWAP-test for different number of shots.

field of quantum tomography [18]. Even on a perfect quantum computer, that does not causes any errors during computing and measurement, every approximation by repeated measurements imposes errors.

Given a reproducible qubit with a two-dimensional unknown wave function $\psi = [a, b]^T$, a measurement in the Pauli-Z direction results in the states $|0\rangle$ and $|1\rangle$ with probabilities $p_0 = |a|^2$ and $p_1 = |b|^2$, respectively. Let us assume we are only interested in the measurement probabilities p_i rather then the values of a and b. We repeat the measurement N times. Every measurement is a random variable $\forall i \in [1, \dots, N] : X_i \overset{iid}{\sim} \text{Bernoulli}(p_0)$.

The value of p_0 can be estimated by the Maximum Likelihood estimation is $\frac{1}{N}\sum_i^N X_i$. The variance of this estimation is $\frac{p_0(1-p_0)}{N}$.

For the SWAP-test the value of p_0 is in the interval $[0, \frac{1}{2}]$ regarding the inner product values in the interval $[0, 1]$. In Fig. 6 the standard deviation regarding the inner product to estimate is depicted. Especially for smaller shot numbers the measurement of inner products, which are close to 0, is impossible. Note that the variance scales by the factor $\frac{1}{N}$, i.e. only linear with N, which implies a high number of shots for higher precision.

This approximation variance yields for both presented strategies. For the SWAP-test we expect a low impact, because the regions of high variance relate to regions of large distances. The smaller distances have a larger effect on the quality and stability of the prototype learning than larger ones. Therefore we expect the SWAP-test to work for LVQ appropriately.

Otherwise, this heterogeneous variance behavior can be a problem when a shifted prototype gets measured. In this case, the distribution of the data set determines the fidelity of the prototype position. Also there is no helpful connection between learning progress and variance, like the SWAP-test, where the variance decreases for closer prototypes in later learning phases. A measurement strategy with more evenly distributed variances is appropriate.

4.2 Conclusion

At this point we determined existing quantum computing patterns that can solve a step of the heuristic LVQ. Combining these steps into larger circuits, that can solve multiple steps in one run will be part of our future work. In this paper, we have addressed the errors that already arise from theoretical limits on a perfect quantum computer. Besides these theoretical limits current quantum computers introduce errors on hardware level, that cannot be compensated yet. The impact of these errors and their possible mitigation will be explored in further work. Combinations of layer patterns from quantum neural networks with the shown patterns might be able to implement relevance prototype learning on a quantum device.

References

1. Aïmeur, E., Brassard, G., Gambs, S.: Quantum clustering algorithms. In: Proceedings of the 24th International Conference on Machine Learning, pp. 1–8. Association for Computing Machinery (2007). https://doi.org/10.1145/1273496.1273497
2. Bausch, J.: Recurrent quantum neural networks. In: Larochelle, H., Ranzato, M., Hadsell, R., Balcan, M.F., Lin, H. (eds.) Advances in Neural Information Processing Systems, vol. 33, pp. 1368–1379. Curran Associates, Inc. (2020). https://papers.nips.cc/paper/2020/file/0ec96be397dd6d3cf2fecb4a2d627c1c-Paper.pdf
3. Biamonte, J., Wittek, P., Pancotti, N., Rebentrost, P., Wiebe, N., Lloyd, S.: Quantum machine learning. Nature **549**, 195–202 (2017). https://doi.org/10.1038/nature23474
4. Feynman, R.: Simulating physics with computers. Int. J. Theor. Phys. **21**, 467–488 (1982). https://doi.org/10.1007/BF02650179
5. Ghosh, S., et al.: Comparison of strategies to learn from imbalanced classes for computer aided diagnosis of inborn steroidogenic disorders. In: Verleysen, M. (ed.) 25th European Symposium on Artificial Neural Networks, Computational Intelligence and Machine Learning, ESANN 2017, pp. 199–205 (2017). https://www.esann.org/sites/default/files/proceedings/legacy/es2017-94.pdf
6. Gitiaux, X., Morris, I., Emelianenko, M., Tian, M.: SWAP test for an arbitrary number of quantum states. arXiv (2021). https://doi.org/10.48550/arXiv.2110.13261
7. Grover, L.K.: A fast quantum mechanical algorithm for database search. In: Proceedings of the Twenty-Eighth Annual ACM Symposium on Theory of Computing, STOC 1996, pp. 212–219. Association for Computing Machinery (1996). https://doi.org/10.1145/237814.237866

8. Hu, W.: Comparison of two quantum nearest neighbor classifiers on IBM's quantum simulator. Nat. Sci. **10**(03), 87–89 (2018). https://doi.org/10.4236/ns.2018.103010
9. Kohonen, T.: Learning vector quantization. Neural Netw. **1**(Suppl. 1), 303 (1988). https://doi.org/10.1016/0893-6080(88)90334-6
10. LaRose, R., Coyle, B.: Robust data encodings for quantum classifiers. Phys. Rev. A **102**(032420) (2020). https://doi.org/10.1103/PhysRevA.102.032420
11. Lengyel, E.: Mathematics for 3D Game Programming and Computer Graphics. Delmar Cengage Learning, Florence (2011)
12. Mengoni, R., Di Pierro, A.: Kernel methods in quantum machine learning. Quantum Mach. Intell. **1**(3), 65–71 (2019). https://doi.org/10.1007/s42484-019-00007-4
13. Rebentrost, P., Bromley, T., Weedbrook, C., Lloyd, S.: Quantum Hopfield neural network. Phys. Rev. A **98**(042308) (2018). https://doi.org/10.1103/PhysRevA.98.042308
14. Rebentrost, P., Mohseni, M., Lloyd, S.: Quantum support vector machines for big data classification. Phys. Rev. Lett. **113**(130503) (2014). https://doi.org/10.1103/PhysRevLett.113.130503
15. Sato, A., Yamada, K.: Generalized learning vector quantization. In: Touretzky, D., Mozer, M.C., Hasselmo, M. (eds.) Advances in Neural Information Processing Systems, vol. 8, pp. 423–429. MIT Press (1995)
16. Schuld, M.: Quantum machine learning models are kernel methods. arXiv (2021). https://doi.org/10.48550/arXiv.2101.11020
17. Schuld, M., Killoran, N.: Quantum machine learning in feature Hilbert spaces. Phys. Rev. Lett. **122**(040504) (2019). https://doi.org/10.1103/PhysRevLett.122.040504
18. Schuld, M., Petruccione, F.: Machine Learning with Quantum Computers. QST, Springer, Cham (2021). https://doi.org/10.1007/978-3-030-83098-4
19. Villmann, T., Engelsberger, A.: Quantum computing for efficient learning in prototype-based vector quantization. In: 26. Interdisziplinäre Wissenschaftliche Konferenz Mittweida, no. 2, pp. 137–140 (2021). https://doi.org/10.48446/opus-12296
20. Villmann, T., Engelsberger, A.: Quantum-hybrid neural vector quantization – a mathematical approach. In: Rutkowski, L., Scherer, R., Korytkowski, M., Pedrycz, W., Tadeusiewicz, R., Zurada, J.M. (eds.) ICAISC 2021. LNCS (LNAI), vol. 12854, pp. 246–257. Springer, Cham (2021). https://doi.org/10.1007/978-3-030-87986-0_22
21. Villmann, T., Engelsberger, A., Ravichandran, J., Villmann, A., Kaden, M.: Quantum-inspired learning vector quantizers for prototype-based classification. Neural Comput. Appl. **34**(1), 79–88 (2020). https://doi.org/10.1007/s00521-020-05517-y
22. Weigold, M., Barzen, J., Leymann, F., Salm, M.: Encoding patterns for quantum algorithms. IET Quantum Commun. **2**(4), 141–152 (2021). https://doi.org/10.1049/qtc2.12032
23. Weigold, M., Barzen, J., Leymann, F., Vietz, D.: Patterns for hybrid quantum algorithms. In: Barzen, J. (ed.) SummerSOC 2021. CCIS, vol. 1429, pp. 34–51. Springer, Cham (2021). https://doi.org/10.1007/978-3-030-87568-8_2
24. Wiebe, N., Kapoor, A., Svore, K.: Quantum algorithms for nearest-neighbor methods for supervised and unsupervised learning. Quantum Inf. Comput. **15**(3 & 4), 316–356 (2015). https://doi.org/10.26421/QIC15.3-4-7

Application of Kohonen Maps in Predicting and Characterizing VAT Fraud in a Sub-Saharan African Country

Ricardo Santos[1](✉), Ricardo Moura[2,3](✉), and Victor Lobo[2,4](✉)

[1] Instituto Superior Técnico, Universidade de Lisboa, 1049-001 Lisbon, Portugal
ricardo.nunes.dos.santos@tecnico.ulisboa.pt, rjgndossantos@gmail.com
[2] Portuguese Navy Research Center, 2810-001 Alfeite, Portugal
[3] Center for Mathematics and Applications, NOVA School of Science and Technology, 2829-516 Caparica, Portugal
[4] NOVA Information Management School, Campus de Campolide, 1070-312 Lisbon, Portugal

Abstract. This paper describes the use Self-Organizing Maps (or Kohonen Networks) in helping to detect Value Added Tax (VAT) fraud. The sub-Saharan African Country presented as a case study has had its largest share in tax revenues since the introduction of VAT, nearly two decades ago, but still has a relatively modest tax efficiency when compared to European or worldwide standards. This trend can be reversed by strengthening the audit and inspection processes of its Revenue Authority (RA) with Data Mining, taking advantage of historical data stored by different information systems. A Case Study in the southern region of this country is presented, where historical data available from tax audits are compared with the VAT returns using Kohonen Maps. Comparing the experimental results with other anomaly detection algorithms, Kohonen maps prove to be of great value in predicting and characterizing VAT fraud in this case study.

Keywords: VAT · Fraud · Self-organizing maps · Audit · Data mining

1 Introduction

VAT constitutes as a barometer [1] to measure a country's Fiscal Gap, which is the estimate resulting from the difference between the volume of expenditure and debt of a country and the volume of its revenue collection over a period of time, assuming negative Fiscal Gap as non-existent [2].

Considered as a very common insufficiency in Sub-Saharan Africa [3], the Fiscal Gap in this part of Africa is, in most of its countries, directly proportional to the VAT fraud rate.

The sub-Saharan African Country (which cannot be disclosed) presented as a case study follows those same lines and its largest share in tax revenues was acquired since the introduction of VAT, nearly two decades ago. Nevertheless,

J. Faigl et al. (Eds.): WSOM+ 2022, LNNS 533, pp. 74–86, 2022.
https://doi.org/10.1007/978-3-031-15444-7_8

these revenues have relatively modest tax efficiency when compared to European or worldwide standards. Its laws are very similar to the European VAT laws, having specific adaptations, driven by a protectionist regime, in particular, for informal economy and tax exemptions in agriculture and fisheries areas. Moreover, VAT collections have greater outputs in the southern region than in other regions, thus leading us to focus this study in that region.

The Revenue Authority (RA) of the this sub-Saharan country has invested, in technological modernization of information systems to interconnect with other relevant systems of the Government and the Private Sector. As a result, greater transparency and better management of all tax processes are being achieved, resulting in a generation of large volumes of structured and unstructured tax data from invoicing systems, particularly those related to VAT. Nevertheless, Fiscal Gap can dramatically be reduced if a more efficient VAT management is implemented.

These are the baselines defined towards the Big Data analysis of the tax ecosystem, in line with the same stance followed by other contemporary tax administrations [5], further strengthening the need for Knowledge Discovery, to build a more assertive Taxpayer profile, despite being a relatively new field of study in African country's public and government sector, in general.

Thus, in the following sections, we present details of the case study and historical data context and compare the data available from tax audits with the VAT returns using Kohonen Maps/Self-Organized Maps (SOM) based algorithms, comparing the experimental results with other anomaly detection algorithms.

2 Case Study

We selected for our case study, VAT data concerning the years 2013–2018. An initial look of the raw data reveals that seven economic sectors, identified here as A, B, C, D, E, F and G due to privacy and confidentiality issues, have been treated differently by Audit, Inspection and VAT Refunds sectors of the RA. Indeed, the focus of audits are taxpayers from sectors B and F which represent the bulk of companies whose capital is mostly owned by national citizens. Most inspections occur on sector B, which is a specific group of taxpayers, representing companies with, transnational/diffuse capital in most of the cases.

Paradoxically, VAT refunds, although keeping a flow of steady payments among all economic sectors mentioned above, disclose a relatively high frequency of beneficiaries in sectors A and E, which are precisely the ones that are normally neglected by audits and inspections. This atypical behaviour suggests a revenue recovery strategy with audits and inspections that is not suited to the public finances expenditure.

Therefore, a case study consisting of the analysis of VAT and related Corporate Income Tax (CIT) returns of 27,049 taxpayers for the years 2013–2018 is conducted to prove the previous statement by answering three (3) hypotheses: (i) the cases under investigation by the Inspection sector correlate with VAT frauds; (ii) refund payments correlate with potential VAT frauds; and (iii) the evaded VAT revenue far exceeds that recovered by the RA.

The sample population of the case study (a sub-saharan country which cannot be dislcosed) are ordinary VAT taxpayers, whose refund is paid in full by the Government and also ordinary VAT taxpayers whose inspections culminate in court cases, excluding those prosecuted by drugs, human trafficking and related offences.

This case study, that might be considered a brand new in the context of this country - and even in some countries whose fiscal reality is similar [6–9], uses data directly collected from the RA information systems and subjected to prior screening, particularly for the treatment of numerous missing fields, which usually characterize VAT and income tax returns.

3 Kohonen Maps

Kohonen Maps [10] or Self-Organized Maps are unsupervised artificial neural networks, which combine multidimensional clustering with visual techniques.

The basic SOM training algorithm can be described as follows: Let X be the set of n training patterns $\mathbf{x}_1, \ldots, \mathbf{x}_n$; W be a $p \times q$ grid of units $\mathbf{w}_{i,j}$ where i and j are their coordinates on that grid; $\alpha \in]0,1[$ be the learning rate initialized to a given initial learning rate; r be the radius of the neighbourhood function $h(\mathbf{w}_{i,j}, \mathbf{w}_{a,b}, r)$, for some $a < p$ and $b < q$ initialized to a given initial radius.

1. Repeat
2. For $k = 1$ to n (*i.e.*, for all $\mathbf{x}_k$)
3. For all $\mathbf{w}_{i,j}$, calculate $d_{i,j} = ||\mathbf{x}_k - \mathbf{w}_{ij}||$
4. Select the unit that minimizes $d_{i,j}$ as the winner $\mathbf{w}_{win(i),win(j)}$
5. Update each $\mathbf{w}_{i,j} \in W$: $\mathbf{w}_{i,j} = \mathbf{w}_{i,j} + \alpha\, h(\mathbf{w}_{i,j}, \mathbf{w}_{win(i),win(j)}, r)||\mathbf{x}_k - \mathbf{w}_{i,j}||$
6. Decrease the value of α and r
7. Until α reaches a given ϵ (near 0)

The neighbourhood function h is usually a function that decreases with the distance (in the output space) to the winning unit, and is responsible for the interactions between different units. During training, the radius r will usually decrease, so that each unit will become more isolated from the effects of its neighbours. The learning rate α must converge to 0 so as to guarantee convergence and stability for the SOM [10].

The good performance of Kohonen Maps is highly influenced by the choice of the number of epochs, neighbourhood function radius and learning rate essentially. Two important metrics contribute to evaluate the quality of the SOM [12,13]: (i) the quantization error (QE) and (ii) the topographical error (TE). The QE shows how closely the SOM units map the input data. High QE values indicate that there is a large difference between the input data and the SOM units, while low values indicate a more perfect match. The lower the QE the better the SOM learning rate, *i.e.*, the clustering. Nevertheless, over-fitting should always be avoided [11]. In turn, the TE metric evaluates the consistency of the mapping, *i.e.*, indicates weather the SOM is folded upon itself in the input space.

The TE counts, amongst all input data, how many times the second best matching unit in the SOM is not adjacent to the best maching unit (BMU). If the TE is high, the SOM is bent over itself in the input space. The lower the TE the better the topographical quality of the SOM, *i.e.*, better visualization/projection of the results. For this reason, it is good practice to train as many Kohonen Maps as possible, using different parametrization, and then choose the one that minimizes QE and TE the most.

The visualization of a Kohonen Map is usually done with the U Matrix (UMAT) which allows to visually identify clusters [14]. Thus, by colour variation, the distance between neurons can be inferred in our software: the distances coded in red tones represent higher values, *i.e.*, more distant neurons, while the distances coded in blue tones represent lower values, *i.e.*, nearby neurons. In turn, greater distances are originated by neurons that incorporate different characteristics of the data, which may correspond to malformed clusters, or data anomalies. On top of a UMAT we may draw a Hits Map that represents the frequency of observations that are grouped in each neuron, thus exposing the distribution of the BMU's. Another useful tool for visualizing Kohonen Maps is the Component Map, which shows the values of each component throughout the SOM. Thus, qualitative information of a sample variable can be obtained about its influence on: (i) formed clusters - a very important aspect to reduce data dimensionality; (ii) correlation and local correlation with the other variables. Kohonen maps have application in various branches of finance [15,16] but also as anomaly detector [17], when combined with k-Means clustering [18] and KNN classification [19] algorithms, originating a robust method whose performance is equivalent to that of other anomaly detection algorithms, *e.g.*, Local Outlier Factor [20] and One-Class SVM [21]. Thus, two variants of the Kohonen Maps are considered the Kohonen QE and Kohonen KNN. Kohonen QE uses the traditional QE measurement approach [13,14]. In this case, the 97th percentile is considered as a threshold. On the other hand, Kohonen KNN [18] previously removes neurons contaminated with noisy data and optimizes QE with the density function of the log-normal distribution $f(x;\mu;\sigma) = \frac{1}{QE\sigma\sqrt{2\pi}} exp\left[-\frac{(\ln(x)-\mu)^2}{2\sigma^2}\right]$. For the Case Study, a 99.7 percentile was assumed as a threshold, meaning that a set of possible fraud neurons are bounded by QE greater than 1.

4 Methodology

To prove the hypotheses presented earlier, a six-stage methodology is observed: (i) extract bulk data from various tax information systems; (ii) data validation using key fields; (iii) tagging of data associated with cases of fraud, investigations and suspicions; (iv) construction of the data cube used in Kohonen Maps; (v) proof of the hypotheses; and (vi) analysis of fraud behaviour patterns.

First, bulk data from: (i) 925,070 ordinary VAT returns (2010–2019); (ii) 452,044 customs declarations (2018–2019); and (iii) 67,554 CIT statements (2015–2019) are cross-checked by valid/existent Taxpayer Identification Numbers (TIN). Since ordinary VAT returns are monthly based and CIT statements annual, VAT returns are summed by fiscal year. Then, cross-check bulk data with samples of audits, inspections and refunds provided by the RA, namely: (i) 131 audits (2013–2018); (ii) 1,050 inspections (2018–2019); and 660 VAT refund requests (2013–2019).

Second, from extracted data generated a valid sample of 27,049 taxpayers under the normal VAT and CIT regime (2013–2018) for the Case Study, representing about 50% of the tax population for that period. Then cross-check TIN, fiscal year and status of compliance to validate samples provided by the RA, which results are: (i) 120 positive cases out of 131 audits[1]; (ii) 127 inspections of taxpayers out 1,050 cases submitted to the Fiscal Court[2]; and (iii) 85 refund payments out of 660 VAT refund requests (see footnote 2).

Third, all samples provided by the RA are marked as fraud (F), for audits with positive cases of fraud; investigations (I), for inspections submitted to the Fiscal Court; and suspicious (S), for VAT refunds actually paid. Since some audits are interspersed in different fiscal years, namely for 2013–2017; 2014–2018; and 2017–2018, redundant data was purged and the clean data was consolidated into a Case Study sample (2013–2018). After, samples of fraud (F), investigations (I) and suspicions (S) are interspersed, by the economic sectors A, B, C, D, E, F and G, among those 27,049 taxpayers, as summarized in Table 1.

Table 1. Frauds, investigations and suspicions (2013–2018) by economic sector

Economic sector	Incidence of fraud	Total of taxpayers	F	I	S
A	Low	594	2	1	23
B	High	10,276	49	93	21
C	Medium-low	3,560	18	4	6
D	Low	3,756	7	5	8
E	Low	117	2	0	9
F	Medium-high	7,981	39	22	13
G	Low	765	3	2	5
Total		27,049	120	127	85

Fourth, a data cube [23] is created with the characteristics: (i) zero value in all unfilled fields; (ii) mean value added to fields with values equal to zero to avoid the division by zero; (iii) fiscal ratios [24,25]; and (iv) Z-Score normalization of fiscal ratios. In the creation of fiscal ratios, the current weighting criteria of regular or unexpected audits by the RA are taken into account: (i) fluctuation in sales volume; (ii) systematic credit; (iii) systematic losses; (iv) requests for refunds; (v) credit offset requests; (vi) anonymous or informant reporting; and (vii) financial behaviour of branches and sectors of activity. As there is no consolidated data on complaints and the financial behaviour of branches and sectors of activity, these are excluded criteria from the Case Study.

[1] Approximately 0.44% of the taxpayers, among which 91.6% are positive cases of fraud, typically unbalanced and biased dataset [29].

[2] Samples from other VAT regimes or with invalid/non-existent TIN are excluded.

Fifth, apply Kohonen Maps based methods to the samples (the fiscal ratios variables are defined as $V1$, $V2$, $V3$, $V4$, $V5$, $V6$, $V7$, $V8$, $V9$, $V10$, $V11$, $V12$, $V13$, $V15$, $V16$, $V18$, $V19$, $V20$ and $V23$, due to privacy matters), and record the frequency of frauds, investigations and suspicions. Validate Hypotheses 1 and 2 in this step. Finally, adopt the Mittal et al. approach [26] for the treatment of truly positive cases[3] of fraud such as: (i) criminal cases submitted to the Public Prosecutor's in 2018 and 2019; and (ii) refund requests deferred and paid from 2014 to 2019. Validate Hypothesis 3 here as follows:

$$R_s = P_{\hat{Y}}.C_s \tag{1}$$

where R_s, $P_{\hat{Y}}$ and C_s are, respectively, the evaded revenue, the precision of the anomaly detection algorithm, and the tax efficiency ratio [22], which stands at 35% [4], *i.e.*, evaded revenue represents close to 2.85 more than the volume of VAT currently submitted by taxpayers.

Sixth, apply Kohonen Components Plan to characterize behavioural pattern of the economic sectors that have historically presented the highest incidence of VAT fraud (see Table 1) using the consolidated map of the Case Study (2013–2018) as a reference. The same procedure applies to the most neglected economic sectors in homologous period. In order to perform the referred stages, we used *Python*, where it must be noted the use of MiniSom[4] library for the training of the SOM maps.

5 Results

Data reveals a strong correlation between some fiscal ratios, which concurs with the literature [15,27]. Consequently, 19 out of 23 initial fiscal ratios are chosen with Decision Trees (CART) variable selectors, and only six (6) with Fisher's Discriminant Analysis. In Table 2, we present the comparative results of the two anomaly detection methods Kohonen QE and Kohonen KNN with the Local Outlier Factor [20] and the One-Class SVM [21]. To handle the unbalanced data when using Local Outlier Factor and One-Class SVM, SMOTE overbalancing technique was applied to the original data.

Table 2. Comparison of the performance of anomaly detection algorithms

Algorithm	Accuracy	Error rate	Recall	Specificity	Precision	F1-Score
Local outlier factor	82.88	17.12	0.22	99.51	8.33	0.46
One-Class SVM	97.26	2.74	7.53	99.75	45.83	0.56
Kohonen QE	96.88	3.12	5.42	99.71	36.67	0.54
Kohonen KNN	99.77	0.23	**85.37**	99.81	**58.33**	**0.85**

[3] States that once registered as a positive case of fraud, the Taxpayer tends to relapse into fraudulent practices in future fiscal years.

[4] https://github.com/JustGlowing/minisom/.

Overall, with 85.37% recall and 58.33% precision, Kohonen KNN, with dimensionality reduced by Fisher's, outperforms the other anomaly detection algorithms used. In any case, the performance of the majority of tested algorithms significantly exceeds the precision achieved by the random method of the RA, which actually corresponds to 120 cases of fraud out of 27,049 taxpayers under the normal VAT regime.

Thus, for the characterization of the frauds, we considered the reduced dimensionality with Decision Trees variable selectors to preserve the majority of fiscal ratios.

It is observed, in Fig. 1, that the UMAT has two distinct hemispheres: (i) the left, where the dark blue outline is visible, which denotes the minimization of the distances of the sample data; and (ii) the right hemisphere where the presence of possible outliers is observed supported by the Hits Map.

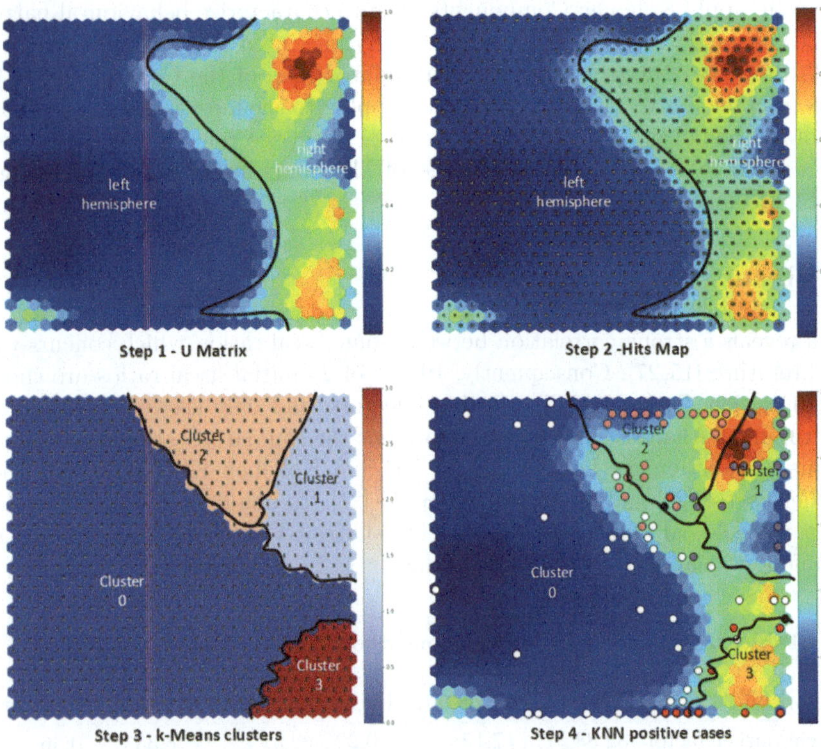

Fig. 1. Summary of the main steps followed in order to make VAT fraud characterization. The coloured dots represent neurons wit positive cases (white - cluster 0; indigo - cluster 1; pink - cluster 2; red - cluster 3), where the black dots represent an superimposition of clusters.

By clustering the SOM neurons with the k-Means, four clusters are obtained, which are topographically distributed over the same hemispheres, in the following order: (i) the left hemisphere, where a well-formed cluster 0 groups neurons with sample data with shorter intra-cluster distances; (ii), in the right hemisphere, clusters 1, 2 and 3 isolate neurons with greater intra-cluster distances, likely outliers or ill-formed clusters. The matching between the positive cases of the Kohonen Map and k-Means clustering (see Fig. 1) is achieved with the KNN classifier [19].

In Fig. 1 - Step 4, the coloured dots show the location of neurons with positive cases isolated by the KNN procedure and some black dots, representing neurons that superimpose to more than one cluster, leading to an undetermined intermediate zone [15]. As exceptional cases, these neurons should be screened by a panel of RA experts.

See also, in Fig. 1, that the topography of the UMAT clusters is not congruent with the one generated by the k-Means, which is explained by the distortion of the map given that, in this case, $TE \approx 0.9$. Thus, by cross-checking the KNN tagged-data with 27,049 taxpayers under the normal VAT regime, the result for each economic sector may be seen in Table 3.

Table 3. Consolidated map of frauds, investigations and suspicions of the case study

Economic sector	Cluster 0			Cluster 1			Cluster 2			Cluster 3		
	F	I	S	F	I	S	F	I	S	F	I	S
A	1	0	6	1	0	2	0	0	1	0	0	0
B	14	24	4	19	1	1	14	3	1	2	3	0
C	4	1	2	0	0	0	4	0	0	10	0	1
D	2	1	1	2	0	0	2	0	0	1	0	0
E	1	0	2	0	0	0	0	0	1	1	0	0
F	16	7	2	5	0	0	9	0	1	9	0	1
G	2	1	0	0	0	0	0	0	0	1	0	0
Total	40	34	17	27	1	3	29	3	4	24	3	2

From Table 3, it is clear that 41 taxpayers are marked as (I) and another 26 as (S), which proves hypotheses 1 and 2. We may also note that economic sectors A and E show a relatively high number of suspects (S) compared to the total

Table 4. Ratio of positive fraud cases by cluster

Cluster	Truly positive cases (F_n)	Frequency of taxpayers (T_n)	Ratio $\left(\frac{F_n}{T_n} \times 10^3\right)$
0	40	7802	5.13
1	27	264	102.27
2	29	564	51.42
3	24	458	52.40

number of frauds (F) reported by the RA and a relatively high fraud detection rate in economic sector B, although accompanied by a significant number of taxpayers marked as investigations (I) and suspects (S). On the other hand, the initial audit sample may be ultimately reduced to 9,088 taxpayers in 27,049, with 66% of positive cases confined to 1,022 taxpayers (clusters 1, 2 and 3) belonging to the right hemisphere, as seen in Table 4.

As F_n is deducted from the frequency of taxpayers in each cluster, T_n, taxpayers grouped in clusters 1, 2 and 3 would be primarily targeted by audits and inspections, in descending order of the ratio $\frac{F_n}{T_n}.10^3$, which can also help prioritize [25] these actions in the field. It is also implied that 0.44% precision achieved by the random method of the RA (refer Table 1) is instantly increased to 6.22%, by simply applying four-step methodology to the Kohonen Map.

Finally, comparing the evaded (R_s) VAT revenue [26] with that recovered (R_r) by the Tax Authority, we have (applying (1) to 58.33% precision of the Kohonen KNN anomaly detector algorithm), in the Case Study, a ratio of 122.09 units, that is, a very significant indicator of VAT tax evasion.

A trend that remains unchanged with a closer look to the economic sectors, seen in Table 5. On the other hand, the behavioural aspects of the perpetrators of VAT fraud, previously mentioned by Pironet [28] and Guarascio [29] are also revealed, with a closer look to the topology of the different multi-coloured points representing real cases of fraud in Fig. 1, which overwhelm converge to the edges of clusters 0, 1, 2 and 3. In order to study the impact of these behavioural aspects, a Components Map for each variable was generated and analysed. and presented in Fig. 2 in Appendix A.

Table 5. Comparison of evaded and recovered revenue by economic sector

Sample	Ratio $\frac{R_s}{R_r}$	Remarks
A	449.67	Cyclically neglected by audits and inspections
B	1304.56	Previously flagged as high fraud incidence
C	75.09	
D	59.17	
E	150.51	Cyclically neglected by audits and inspections
F	8.22	Previously flagged as medium-high fraud incidence
G	1444.41	

It follows that the recommended audit strategy for the neglected economic sectors A (clusters 0, 1 and 2) and E (clusters 0, 2 and 3) could be as it is described in Table 6.

In short, different audit and inspection strategies should be put in place for economic sectors A and E considering VAT credits and Income Tax Losses indicators.

Table 6. Audit and inspection strategy induced by case study components map

Sector	Indicator	Priority (Taxpayers under the normal VAT regime)
A	Tax base	With below-average operations and deductions
	VAT credits	With regime with below-average reported credits
	VAT return	With below-average returns
	Corporate income tax losses	Who report income tax with below-average losses.
E	Tax Base	With below-average operations and deductions
	VAT credits	With any reported credits
	VAT return	With below-average returns
	Corporate income tax losses	With any income tax losses

6 Conclusions

With the application of Kohonen Maps, we managed to achieve 58.33% fraud detection rate on the case study data, far exceeding the 0.44% fraud detection rate obtained with the random method currently used by the RA. In addition, by using Kohonen Maps we can drastically reduce the initial sample of taxpayers, with 66% of the spotted positive cases coming from three identified clusters, showing the relevance of using Kohonen maps for prioritizing audits and inspections and improving the detection rate.

In this study, we confirmed that cases under investigation by the Inspection sector correlate with VAT frauds and that refund payments correlate with potential VAT frauds. With a precision of 58.33%, we confirmed that the evaded VAT revenue far exceeds that recovered by the RA, showing that recovered revenue by RA is far below the evaded one. This trend does not change even when analysing by the different economic sectors, where the dis-proportionality between the volume of audits, inspections and the payments due to VAT refunds becomes clear. In short, the current revenue recovery strategy can be greatly improved using our methodology.

It is also clear that economic sectors with higher incidence of VAT fraud, the allocation of human, financial or material resources by the RA in audits and inspections does not have a deterrent effect on potential fraudsters. On the other hand, among neglected economic sectors, the proportion of evaded revenue exceeds in magnitude that of the economic sectors with medium and high incidence of VAT fraud. Finally, Components Map outlines better auditing and inspection strategies aimed at economic sectors cyclically neglected by the audits and inspections of the Tax Authority.

As future studies, it is intended to analyse the impact of incorporating an Auto-Encoder into the model proposed here, to predict and characterize VAT fraud in southern region, namely, to deal with semi-structured and unstructured data generated by the interoperability that exists between the information systems of RA and banking, public procurement, civil identification, commercial registration, and social security counterparts, among others.

Appendix A - Component Maps

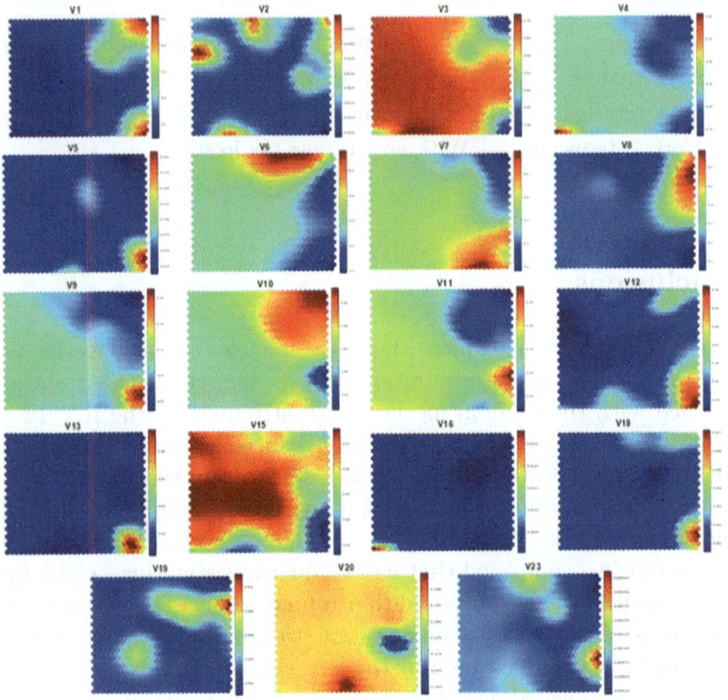

Fig. 2. Case study components maps.

References

1. Hajdúchová, I., sedltačiková, M., Viszlat, I.: Value-added tax impact on the state budget expenditures and incomes. Procedia Econ. Finance, **34**, 676–681 (2015)
2. Duda, R.O., Hart, P.E., Stork, D.G.: Pattern Classification, 2nd edn., pp. 1–19. Wiley, Hoboken (2000)
3. Krever, R.: VAT in Africa, pp. 71–80. Pretoria University Law Press, Pretoria (2008)
4. La Feria, R., Schoeman, A.: Addressing VAT fraud in developing countries: the tax policy-administration symbiosis. Intertax **47**(11), 950–967 (2019)
5. IMF: Digital Revolutions in Public Finance (2017). https://doi.org/10.5089/9781484315224.071. Accessed 25 May 2021

6. Mwanza, M., Phiri, J.: Fraud detection on bulk tax data using business intelligence data mining tool: a case of Zambia revenue authority. Int. J. Adv. Res. Comput. Commun. Eng. **5**(3), 793–798 (2016)
7. De Roux, D., Pérez, B., Moreno, A., Villamtl, M.D.P., Figueroa, C.: Tax fraud detection for under-reporting declarations using an unsupervised machine learning approach. In: KDD, 19–23 August 2018, London, UK, pp. 215–222 (2018)
8. Jihal, H., Talhaou, M.A., Daif, A., Azzouazi, M.: Predictive analytics as a service on moroccan tax evasion. Int. J. Eng. Technol. **7**(432), 90–92 (2018)
9. Jupri, M., Sarno, R.: Data mining, fuzzy AHP and TOPSIS for optimizing taxpayer supervision. Indonesian J. Electr. Eng. Comput. Sci. **18**(1), 75–87 (2020)
10. Kohonen, T.: Self-organized formation of topologically correct feature maps. Biol. Cybern. **43**(1), 59–69 (1982). https://doi.org/10.1007/BF00337288
11. Wendel, J., Buttenfield, B.: Formalizing guidelines for building meaningful self-organizing maps. In: GIScience 2010: Sixth international conference on Geographic Information Science. Zurich, 14-17th September (2010)
12. Tu, L.A., Thai, V.D., Hoan, N.Q.: Improving feature map quality of SOM based on adjusting the neighborhood function. Int. J. Comput. Sci. Inf. Secur. (IJCSIS) **14**(9), 1–6 (2016)
13. Kohonen, T., Schroeder, M.R., Huang, T.S.: Self-Organizing Maps, 3rd edn. Springer-Verlag, Berlin, Heidelberg (2001). https://doi.org/10.1007/978-3-642-56927-2
14. Ultsch, A: U*-Matrix: a Tool to visualize Clusters in high dimensional Data. DataBionics Research Lab, Department of Computer Science University of Marburg, Technical report no. 36 (2003)
15. Serrano-Cinca C.: Let Financial Data Speak for Themselves. In: Deboeck, G., Kohonen, T. (eds.) Visual Explorations in Finance with Self-Organizing Maps. Springer Finance. Springer, London. pp. 3–18 (1998). https://doi.org/10.1007/978-1-4471-3913-3_1
16. Kiviluoto, K., Bergius, P.: Maps for analyzing failures of small and medium-sized enterprises. In: Deboeck, G., Kohonen, T. (eds.) Visual Explorations in Finance with Self-Organizing Maps. Springer Finance. Springer, London. pp. 59-71 (1998). https://doi.org/10.1007/978-1-4471-3913-3_4
17. Tian, J., Azarian, M.H., Pecht, M.: Anomaly detection using self-organizing maps-based k-nearest neighbor algorithm. In: PHM Society European Conference, vol. 2, no. 1, pp. 3–4 (2014)
18. Steinhaus, H.: Sur la division des corps matériels en parties. Bull. Acad. Polon. Sci. **4**(12), 801–804 (1957)
19. Fix, E., Hodges, J.L.: Discriminatory Analysis. Consistency Properties. USAF School of Aviation Medicine, Randolph Field, Texas, Nonparametric Discrimination (1951)
20. Breunig, M.M., Kriegel, H., NG, R.T., Sander, J. : LOF: identifying density-based local outliers. In: SIGMOD Rec, vol. 29, 2 June, pp. 93-104 (2000)
21. Schölkopf, B., Williamson, R.C., Smola, A.J., Shawe-Tay, J., PLATT, J.C.: Support vector method for novelty detection. In: NIPS 1999 Proceedings of the 12th International Conference on Neural Information Processing Systems, November, pp. 582–588 (1999)
22. Cnossen, S.: Mobilizing VAT revenues in African countries. Int. Tax Public Fin. **22**(6), 1077–1108 (2015). https://doi.org/10.1007/s10797-015-9348-1
23. Wu, R., Ou, C.S., Lin, H., Chang, S., Yen, D.C.: Using data mining technique to enhance tax evasion detection performance. Expert Syst. Appl. **39**, 8769–8777 (2012)

24. Basta, S., et al.: High quality true-positive prediction for fiscal fraud detection. In: 2009 IEEE International Conference on Data Mining Workshops (2009)
25. Vanhoeyveld, J., Martens, D., Peeters, B.: Value-added tax fraud detection with scalable anomaly detection techniques. Appl. Soft Comput. J. **8640**, 1–20 (2019)
26. Mittal, S., Reich, O., Mahajan, A.: Who is bogus? using one-sided labels to identify fraudulent firms from tax returns. In: Proceedings of the 1st ACM SIGCAS Conference on Computing and Sustainable Societies, in COMPASS 2018. ACM, New York, NY, USA, pp. 24:1–24:11 (2018)
27. Matos, T., Macedo, J.A.F., Monteiro, J.M.: An empirical method for discovering tax fraudsters: a real case study of Brazilian fiscal evasion. In: Proceedings of the 19th International Database Engineering No. 38, Applications Symposium, in IDEAS 2015. ACM, New York, NY, USA, pp. 41-48 (2015)
28. Pironet, M.: Classification for Fraud Detection with Social Network Analysis. Master's Thesis, Instituto Superior Técnico. Department of Computer Engineering. Master in Informatics and Computer Engineering (2009)
29. Guarascio, M.: Data Mining Techniques for Fraud Detection. Ph. D thesis, Universitádella Calabria. Dipartimento di Elletronica, Informatica e Sistemistica. Dottorato di Ricerca in Ingegneria dei Sistemi e Informatica. Ciclo XXII (2010)

GNG-based Clustering of Risk-aware Trajectories into Safe Corridors

Jakub Sláma(✉), Petr Váňa, and Jan Faigl

Department of Computer Science, Faculty of Electrical Engineering, Czech Technical University in Prague, Technická 2, 166 27 Prague 6, Czech Republic
{slamajak,vanapet1,faiglj}@fel.cvut.cz
https://comrob.fel.cvut.cz/

Abstract. Personal air transportation on short distances is a promising trend in modern aviation, raising new challenges as flying in low altitudes in highly populated environments induces additional risk to people and properties on the ground. Risk-aware planning can mitigate the risk by preferring flying above low-risk areas such as rivers or brownfields. Finding such trajectories is computationally demanding, but they can be precomputed for areas that are not changing rapidly and form a planning roadmap. The roadmap can be utilized for multi-query trajectory planning using graph-based search. However, a quality roadmap is required to provide a low-risk trajectory for an arbitrary query on a risk-aware trajectory from one location to another. Even though a dense roadmap can achieve the quality, it would be computationally demanding. Therefore, we propose to cluster the found trajectories and create a sparse roadmap of safe corridors that provide similar quality of risk-aware trajectories. In this paper, we report on applying Growing Neural Gas (GNG) in estimating the suitable number of clusters. Based on the empirical evaluation using a realistic urban scenario, the results suggest a significant reduction of the computational burden on risk-aware trajectory planning using the roadmap with the clustered safe corridors.

Keywords: Growing neural gas · Risk-aware planning · Urban air mobility

1 Introduction

Urban Air Mobility (UAM) is an emerging field in the aerospace industry aiming to provide personal air transportation on short distances [6] that would result in raising the number of small aircraft flying in urban areas [8], which will increase the risk of possible accidents. Any accident is not only a threat to the aircraft and people on board but also to people and properties on the ground. A high population density can characterize urban areas, and so any crash can have immense consequences, such as the number of casualties or the caused material damage. Hence, the UAM brings challenges in risk mitigation.

J. Faigl et al. (Eds.): WSOM+ 2022, LNNS 533, pp. 87–97, 2022.
https://doi.org/10.1007/978-3-031-15444-7_9

The risk can be mitigated by risk-aware trajectory planning [10] to find the least risky trajectory with the minimal induced risk in the case of a malfunction. Risk evaluation can be prohibitively demanding for real-time queries or simultaneous planning of multiple aircraft. Therefore, a roadmap of risk-aware trajectories can be determined to support graph-based planning methods. Furthermore, the found least risky trajectories tend to be over less populated areas such as rivers or less risky brownfields [7]. Such low-risk trajectory segments can be called safe corridors, and similar safe corridor segments are common among different least risky trajectories. Hence, the safe corridors can be determined and utilized to determine a roadmap that would be a relatively dense roadmap of safe (individual) trajectories otherwise. Then, existing graph-based planning techniques can be employed for fast near-optimal risk-aware trajectory planning.

In this paper, we propose a clustering method based on the *Growing Neural Gas* (GNG) algorithm [5] to determine safe corridors from a determined set of least risky trajectory samples. The problem is determining a relatively spare roadmap that would provide risk-aware trajectories with similar risk to the original set of trajectories. The roadmap of the safe corridors can then be employed in computationally efficient multi-query risk-aware trajectory planning. The approach is based on the unsupervised clustering of the GNG to determine the roadmap vertices in the most used locations in the set of input trajectories. The roadmap edges are then determined based on relations between the vertices encoded in the GNG structure.

2 Problem Specification

The addressed problem of determining safe corridors from the set of available risk-aware trajectories builds on the existing work to determine the least risky trajectories [11]. However, the formal vehicle model is needed to determine the final trajectory roadmap because it induces the proposed distances between two sampled trajectory segments required in the clustering. Therefore, a brief overview of the notion is presented to make the paper self-contained. For a detailed definition of the least risky trajectory planning problem, the reader is referred to [11].

The aircraft is modeled as Dubins Airplane [2] with the configuration $q = (x, y, z, \theta, \psi)$ with the position $(x, y, z) \in \mathbb{R}^3$, $\theta \in \mathbb{S}$ denoting the heading angle, and $\psi \in \mathbb{S}$ standing for the pitch angle. The trajectories are thus determined in the configuration space $\mathcal{C} = \mathbb{R}^3 \times \mathbb{S}^2$ and the aircraft state can be expressed as

$$\begin{bmatrix} \dot{x} \\ \dot{y} \\ \dot{z} \\ \dot{\theta} \end{bmatrix} = v \begin{bmatrix} \cos\theta\cos\psi \\ \sin\theta\cos\psi \\ \sin\psi \\ u_\theta \rho^{-1} \end{bmatrix}, \tag{1}$$

where v is the forward velocity of the aircraft, $u_\theta \in [-1, 1]$ denotes the control input controlling the heading angle θ, and ρ is the minimum turning radius of the aircraft. The pitch angle ψ is considered to change significantly faster than

θ, and thus abrupt changes in ψ are allowed, but $\psi \in [\psi_{\min}, \psi_{\max}]$ has to be satisfied. Trajectories are planned in the collision-free part of $\mathcal{C}$ denoted as $\mathcal{C}_{\text{free}}$.

For a trajectory $\Gamma : [0, T_\Gamma] \to \mathcal{C}_{\text{free}}$ going from an initial configuration $q_i = \Gamma(0)$ to the final configuration $q_f = \Gamma(T_\Gamma)$, the risk r induced by Γ is given as

$$r(\Gamma) = \int_0^{T_\Gamma} \mathcal{M}(\Gamma(t))\, \mathrm{d}t\,, \tag{2}$$

where $\mathcal{M}$ is an aircraft specific risk map, and $\mathcal{M}(q)$ denotes the risk associated with any configuration q. Note that $\Gamma(t)$ denotes the aircraft configuration q on the trajectory Γ at the time t. Planning the least risky trajectory Γ^* between two configurations q_i and q_f stands to find a feasible trajectory from q_i to q_f with the minimal induced risk r that can be expressed as

$$\Gamma^* = \arg\min_{\Gamma} r(\Gamma)\,, \qquad \text{s.t.} \quad \Gamma^*(0) = q_i,\; \Gamma^*(T_{\Gamma^*}) = q_f\,, \tag{3}$$

with the associated risk

$$r^*(q_i, q_f) = r(\Gamma^*)\,. \tag{4}$$

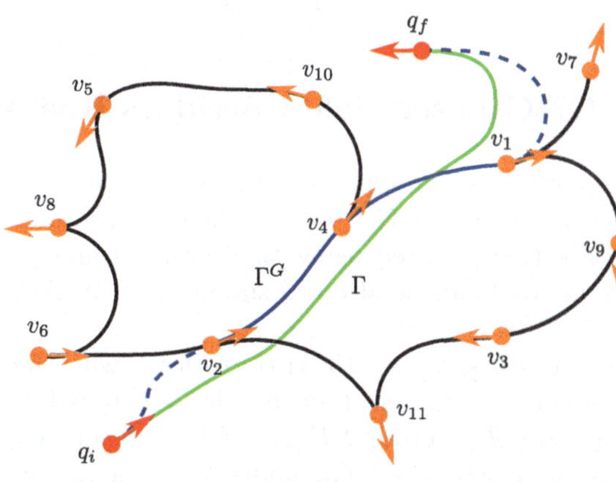

Fig. 1. An example of a safe corridors roadmap G with vertices $\boldsymbol{V}$ (in orange). The risk-aware trajectory Γ (in green) between configurations q_i and q_f (in red) can be approximated by the trajectory Γ^G (in blue) found through the roadmap G. The configurations q_i and q_f are inserted into the roadmap (shown by dashed lines), and the least risky trajectory in the graph is found. As Γ^G is only an approximation of Γ, its risk can be higher compared to the risk of Γ.

Denote $\Xi = \{\Gamma_1, \Gamma_2, \ldots, \Gamma_n\}$ as a set of least risky trajectories from various locations covering the operational area. Let $G(\boldsymbol{V}, \boldsymbol{E})$ denote a roadmap of trajectories connecting configurations associated to the vertices $\boldsymbol{V}$ connected by edges $\boldsymbol{E}$ representing the least risky trajectories. Since G created from Ξ would be very dense and thus graph-based searching for the least risky trajectory would be demanding, the problem is to find G representing trajectories Ξ such that finding an approximation of the least-risky trajectory Γ^G using the roadmap G would be less demanding than determining Γ. The aim is to find G that minimizes the relative risk between trajectories $\Gamma_i \in \Xi$ and their corresponding approximations from G. An example of trajectories is depicted in Fig. 1.

Let Γ^G be the least risky trajectory from q_i to q_f through G. Γ^G visits the intermediate configurations $v_i \in \boldsymbol{V}$ that can be expressed as a sequence of indices $\Sigma = \{\sigma_1, \sigma_2, \ldots, \sigma_{|\Sigma|}\}$. The corresponding risk r^G of the roadmap-based trajectory Γ^G can be expressed as

$$r^G(q_i, q_f) = \min_{\Sigma} \left[r(q_i, v_{\sigma_1}) + \sum_{k=1}^{|\Sigma|-1} r(v_{\sigma_k}, v_{\sigma_{k+1}}) + r(v_{\sigma_{|\Sigma|}}, q_f) \right]. \quad (5)$$

The addressed creation of the safe corridors roadmap G is a problem determining G with a suitable number of vertices k from the trajectories Ξ such that the mean relative risk between the original trajectory Γ and its approximation Γ^G is minimal. The optimization problem can be formally defined as Problem 1.

Problem 1 (Determining Safe Corridors Roadmap G).

$$\underset{G}{\text{minimize}} \ \frac{1}{n} \sum_{j=1}^{n} \frac{r^G(q_i^j, q_f^j)}{r(\Gamma_j)} \quad (6)$$

3 Proposed GNG-based Safe Corridors Roadmap

The proposed safe corridors roadmap determination is based on clustering the given set of trajectories Ξ using GNG to learn the best locations for the roadmap nodes. The edges are then created using the learned topology encoded in the GNG. The proposed roadmap creation is summarized in Algorithm 1, and it works as follows.

The algorithm starts by sampling the trajectories Ξ with the sampling step Δ (Lines 3 to 5, Algorithm 1); the set of all samples is denoted as $\mathcal{Q}$. The samples are from the *Special Euclidean* group $SE(2) = \mathbb{R}^2 \times \mathbb{S}$, where a sample is given by its position and also its orientation. The nodes $\boldsymbol{V}$ of the safe corridors roadmap G are found by the GNG [5] modified for the $SE(2)$ configuration space by the proposed `LearnGNG` method detailed in Sect. 3.1. Once the GNG is terminated, the found nodes serve as the nodes of the safe corridors roadmap G (Line 7, Algorithm 1). Then, the edges of G are created using GNG edges that are non-directional and only denote the relations between nodes, not the actual corridors. Therefore a direction with the shorter feasible maneuver is preferred as the directed edge of the roadmap G (Lines 8 to 12, Algorithm 1).

3.1 GNG-based Safe Corridors Topology Learning

The original GNG algorithm [5] is adapted for $SE(2)$ to respect a distance between two nodes, which depends on the Euclidean distance and also on the

Algorithm 1: Safe Corridors Roadmap Creation

Input: $\mathcal{M}$ – Risk map; Ξ – Set of trajectories.
Parameters: Δ – Sampling step; $\mathcal{A}$ – Aircraft model.
Output: G – Roadmap of safe corridors.

Function `CreateCorridors`($\mathcal{M}, \Xi$):
  $\mathcal{Q} \leftarrow \emptyset$
  forall $\Gamma_j \in \Xi$ **do**
    $\{q_1^j, q_2^j, \ldots, q_{m_j}^j\} \leftarrow$ `SamplePath`(Γ_j, Δ)
    $\mathcal{Q} \leftarrow \mathcal{Q} \cup \{q_1^j, q_2^j, \ldots, q_{m_j}^j\}$
  $G_{\text{GNG}} \leftarrow$ `LearnGNG`$(\mathcal{Q})$
  $G \leftarrow \{\boldsymbol{V} \leftarrow \boldsymbol{V}_{\text{GNG}}, \boldsymbol{E} \leftarrow \emptyset\}$
  forall $(s_1, s_2) \in \boldsymbol{E}_{\text{GNG}}$ **do**
    if $\mathcal{L}$(`Dubins3D`$(s_1, s_2, \mathcal{A})$) $<$ $\mathcal{L}$(`Dubins3D`$(s_2, s_1, \mathcal{A})$) **then**
      $\boldsymbol{E} \leftarrow \boldsymbol{E} \cup \{(s_1, s_2)\}$
    else
      $\boldsymbol{E} \leftarrow \boldsymbol{E} \cup \{(s_2, s_1)\}$
  return G

vehicle's heading angle. The algorithm is summarized in Algorithm 2, and it works as follows.

The GNG constructs a group roadmap G_{GNG} initialized by two nodes s_a and s_b at randomized locations q_a and q_b (Lines 2 to 4, Algorithm 2). Then, the learning epoch is repeated until a termination condition is met. In each iteration, a random configuration (signal) is generated (Line 6, Algorithm 2) with the probability distribution that should correspond to the probability distribution of the underlying data; in our case, the underlying topology of safe corridors. Since the trajectories Ξ are sampled uniformly, the condition is satisfied by selecting random samples from $\mathcal{Q}$.

In the second step, the closest node s_1 and the second-closest node s_2 to ξ are found by the `GetTwoClosestNodes` method. The distance function `Dist` between two configurations q_1 and q_2 is computed as

$$\texttt{Dist}(q_1, q_2) = \max\left(\left\| q_1^{2\text{D}} - q_2^{2\text{D}} \right\|, \frac{d_{\max}}{\Delta_{\max}^{\theta}} \left| \angle(q_1^{\theta}, q_2^{\theta}) \right| \right), \tag{7}$$

where $\left\| q_1^{2\text{D}} - q_2^{2\text{D}} \right\|$ is the Euclidean distance between the samples, $\angle(q_1^{\theta}, q_2^{\theta})$ is the angular difference between their headings, and $\Delta_{\max}^{\theta}$ and $d_{\max}$ are the maximum heading difference and Euclidean distance between two samples to be considered as near, respectively. Thus, the closest node s_1 and the second closest node s_2 are given as

$$s_1 = \arg\min_{v_i \in \boldsymbol{V}} \texttt{Dist}(\texttt{q}(v_i), \xi), \tag{8}$$

$$s_2 = \arg\min_{v_i \in \boldsymbol{V} \setminus s_1} \texttt{Dist}(\texttt{q}(v_i), \xi), \tag{9}$$

where $\texttt{q}(\cdot)$ provides the configuration q associated with the given node.

Algorithm 2: Learning of GNG (adopted from [5])

Input: $\mathcal{Q}$ – Set of samples.

Parameters: ϵ_b and ϵ_n – Attraction parameters for the closest node and neighbors of the closest node, respectively; $a_{\max}$ – Maximum edge age; λ – Number of generated signals before inserting new node; α and d – Error reduction parameters when inserting new node and for all nodes, respectively.

Output: G_{GNG} – GNG Roadmap.

1 **Function** `LearnGNG`($\mathcal{Q}$):
  `/* GNG Initialization */`
2   $q_a, q_b \leftarrow$ `RandomConfigurations`($\mathcal{M}$)
3   $s_a, s_b \leftarrow$ `CreateNodes`(q_a, q_b)
4   $G_{\text{GNG}} \leftarrow \{\boldsymbol{V}_{\text{GNG}} \leftarrow \{s_a, s_b\}, \boldsymbol{E}_{\text{GNG}} \leftarrow \{(s_a, s_b)\}\}$
  `/* Learning phase of GNG */`
5   **while** *Terminal condition is not met* **do**
6     $\xi \leftarrow$ `Random`($\mathcal{Q}$); $n_\xi \leftarrow n_\xi + 1$ `// Step 1: signal generation`
7     $s_1, s_2 \leftarrow$ `GetTwoClosestNodes`($\boldsymbol{V}_{\text{GNG}}, \xi$) `// Step 2: nearest nodes`
8   **forall** $e \in$ `GetConnections`(G_{GNG}, s_1) **do** `// Step 3: age update`
9   `Age`(e) $\leftarrow$ `Age`(e) $+ 1$
10   `Error`(s_1) $\leftarrow$ `Error`(s_1) $+$ `Dist`(s_1, ξ) `// Step 4: error update`
11   `q`(s_1) $\leftarrow$ `q`(s_1) $+ \epsilon_b\,(\xi -$ `q`(s_1)$)$ `// Step 5: position update`
12   **forall** $n \in$ `GetNeighbors`(G_{GNG}, s_1) **do**
13   `q`(n) $\leftarrow$ `q`(n) $+ \epsilon_n\,(\xi -$ `q`(s_1)$)$
14   **if** $(s_1, s_2) \in \boldsymbol{E}_{\text{GNG}}$ **then** `// Step 6: edge update`
15   `Age`$((s_1, s_2)) \leftarrow 0$
16   **else**
17       $\boldsymbol{E}_{\text{GNG}} \leftarrow \boldsymbol{E}_{\text{GNG}} \cup (s_1, s_2)$
18   **forall** $e \in \boldsymbol{E}_{\text{GNG}}$: `Age`(e) $> a_{\max}$ **do** `// Step 7: edge removal`
19       $\boldsymbol{E}_{\text{GNG}} \leftarrow \boldsymbol{E}_{\text{GNG}} \setminus e$
20   **if** $n_\xi > \lambda$ **then** `// Step 8: node insertion`
21       $n_\xi \leftarrow 0$
22   `InsertNewNode`(G_{GNG})
23   **forall** $n \in \boldsymbol{V}$ **do** `// Step 9: error reduction`
24   `Error`(n) $\leftarrow d\,$`Error`(n)
25   **return** G_{GNG}

For the position update (Lines 11, Algorithm 2) a vector from s_1 to ξ is calculated, where we exploit features of Euler angles and the heading difference is calculated as

$$\theta_\xi - \theta_{s_1} = ((\theta_\xi - \theta_{s_1} + \pi) \bmod 2\pi) + \pi \tag{10}$$

to get a correctly oriented angle difference. The configuration of s_1 is adapted towards ξ with the power of ϵ_b (Line 11, Algorithm 2), and all direct neighbors of s_1 are adapted towards ξ with the power of ϵ_n (Lines 12 and 13, Algorithm 2).

The further important modification of the GNG is the node insertion. If λ random configurations (signals) ξ have been generated since the last node insertion, a new node is generated by the `InsertNewNode` method and inserted into G_{GNG} (Lines 20 to 22, Algorithm 2). The insertion is summarized in Algorithm 3.

Algorithm 3: Insertion of a new node into the GNG roadmap

Input: $G_{\mathrm{GNG}}(\boldsymbol{V}, \boldsymbol{E})$ – GNG Roadmap.
Output: Updated roadmap $G_{\mathrm{GNG}}(\boldsymbol{V}, \boldsymbol{E})$.

Function `InsertNewNode`(G_{GNG}):

$q \leftarrow \arg\min_{v_i \in V} \mathtt{Error}(v_i)$

$f \leftarrow \arg\min_{v_i \in \mathtt{GetNeighbors}(G_{\mathrm{GNG}}, q)} \mathtt{Error}(v_i)$

$r \leftarrow \mathtt{NewNode}\left(\frac{\mathtt{q}(q)+\mathtt{q}(f)}{2}\right)$

$\boldsymbol{E} \leftarrow \boldsymbol{E} \setminus (q, f) \cup (q, r) \cup (r, f)$

$\mathtt{Error}(q) \leftarrow \alpha\, \mathtt{Error}(q)$

$\mathtt{Error}(f) \leftarrow \alpha\, \mathtt{Error}(f)$

$\mathtt{Error}(r) \leftarrow \mathtt{Error}(q)$

During the insertion, the node q with the maximal error and its neighbor f with the maximal error are determined. A new node r with the configuration corresponding to the midpoint between q and f is created and connected to q and f while the edge between q and f is removed. Finally, the errors of q and f are reduced by the factor α, and the error of r is assumed to be identical to the error of q.

4 Results

The proposed method has been empirically evaluated in a realistic urban scenario to test its performance and behavior. The test scenario consists of 5 km $\times$ 5 km large urban area of the Prague city center adopted from [11]; the scenario is visualized in Fig. 2. The herein considered risk is measured as on-ground casualties in the case of a crash, and so the risk map is mainly given by the population density and shelter provided by the environment. The utilized aircraft model corresponds to Cessna 172 with the minimum turning radius $\rho = 65.7\,\mathrm{m}$ adopted from [12]. The set Ξ of 3 678 trajectories between randomly generated pairs of start and goal configurations has been found by the RRT*-based risk-aware trajectory algorithm [11]. The algorithm has been run for 750 s for each trajectory. Trajectories are visualized in Figs. 3a to 3b.

The maximum Euclidean distance $d_{\mathrm{max}} = 250\,\mathrm{m}$ and maximum angular difference $\Delta^{\theta}_{\mathrm{max}} = 30°$ have been used in the distance function (7). The used GNG learning parameters are $\epsilon_b = 0.3$, $\epsilon_n = 0.01$, $\alpha = 0.75$, $d = 0.995$, the maximum allowed edge age $a_{\mathrm{max}} = 300$, and a new node is inserted after generating $\lambda = 300$ signals. The utilized terminal condition is the number of vertices k

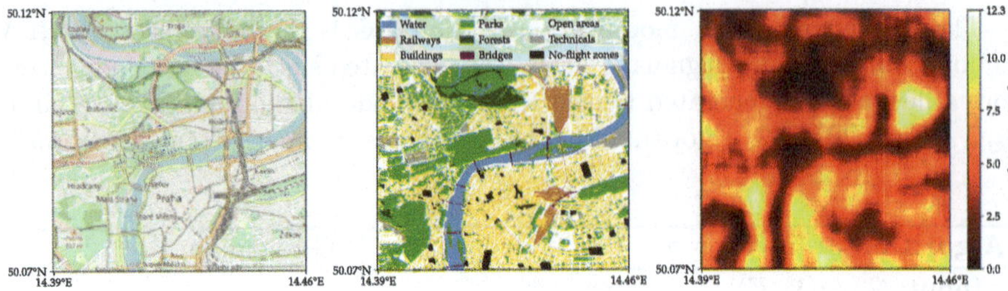

(a) Map (b) Area type (c) Population density

Fig. 2. Selected layers of the utilized urban scenario based on the Prague city center adopted from [11]. The map layer is obtained from the OpenStreetMap [9] and real population density layer is based on [4].

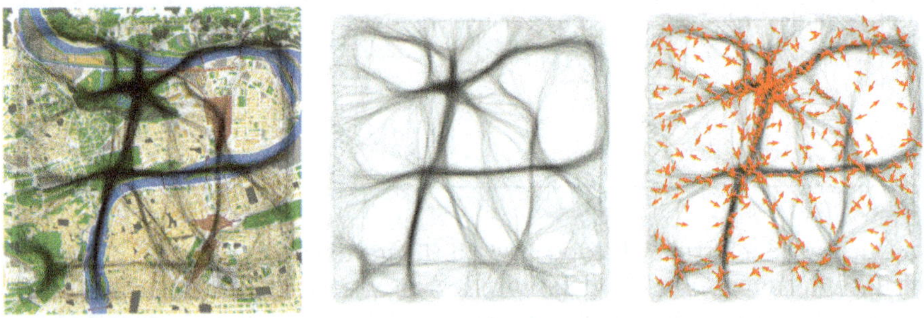

(a) Trajectories over the city map (b) Heatmap of the least risky trajectories (c) Roadmap vertices

Fig. 3. A heatmap of the least risky trajectories obtained by [11] over (**a**) the city map, (**b**) the resulting heatmap of 3 678 trajectories, and (**c**) 500 vertices of the GNG-based roadmap (orange). Darker areas mean more trajectories passing through them.

selected as $k \in \{25, 50, 100, 250, 500, 1\,500\}$. The risk-based planning algorithm from [11] has been used for the evaluation of risk associated with the determined safe corridors. The method has been implemented in Julia ver. 1.6.2 [1] and executed on a single core of the Intel© Core™ i7-9700 CPU. An example of the found safe corridor roadmap's nodes for $k = 500$ is depicted in Fig. 3c.

The quality of the roadmap G is evaluated using the risk induced by the trajectories between the initial and final configurations of the trajectories Ξ. The configurations have been inserted into the found roadmap, and the least risky path has been extracted using Dijkstra's algorithm [3]. The quality of G is characterized by the ratio of the sum of risks induced by all the trajectories Ξ.

Table 1. Roadmap quality given by (6) based on 3 678 trajectories

No. of vertices k	25	50	100	250	500	1 500
Roadmap quality [-]	1.80	1.33	1.27	1.08	1.08	1.01

Table 2. Computational demands of the proposed method

No. of vertices	25	50	100	250	500	1 500
GNG Creation [s]	2	5	19	117	353	3 452
Roadmap edges creation [s]	6	11	18	33	47	103
Avg. query time [s]	2.3	1.9	1.8	1.3	1.0	0.8

The risk of the original trajectories and trajectories found using the safe corridors roadmap G has been evaluated using [11] and the results are depicted in Fig. 4. The quality of the roadmap is measured according to (6) with lower values meaning the higher quality of the roadmap. The roadmap quality for the evaluated number of vertices k is listed in Table 1, and the computational requirements are in Table 2.

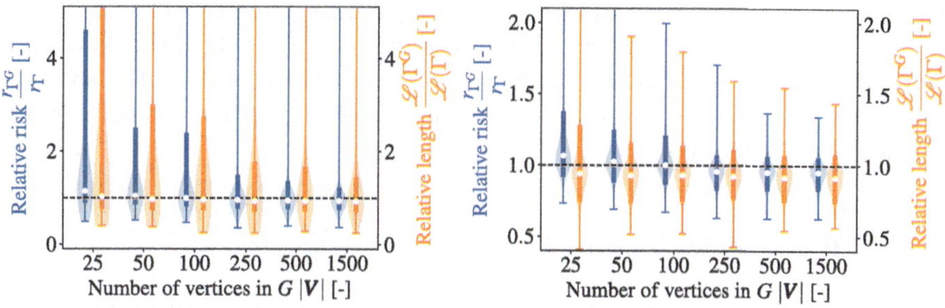

(a) All scenarios

(b) Scenarios where $||q_i^{2D} - q_f^{2D}|| > 2000\,\mathrm{m}$

Fig. 4. Influence of the number of roadmap vertices k on the trajectory relative risk and relative length compared to the original trajectories Ξ: (**a**) all solved scenarios, (**b**) scenarios where q_i and q_f are further apart. Medians (white dots) and 90% non-parametric intervals (thick vertical lines) are shown. The relative risk and relative length reduce with the increasing number of the roadmap vertices as the found safe corridors provide better coverage.

The results support the idea that the safe corridors roadmap can provide trajectories with similar induced risk to the original trajectories Ξ. Increasing k increases the roadmap density that better covers the area. Notice the query times are decreased with increasing k. It is because the risk of maneuvers inserting initial and final configurations q_i and q_f into the roadmap needs to be evaluated, and the maneuvers are generally shorter with increasing roadmap density; the utilized risk evaluation [11] scales with the maneuver length. The best roadmap with $k = 1\,500$ vertices is further analyzed.

For $k = 1\,500$, the median risk reduction is 4% compared to the reference method represented by the trajectories Ξ. The 90% non-parametric interval of the relative risk is [0.77, 1.25]. Furthermore, the median length reduction is

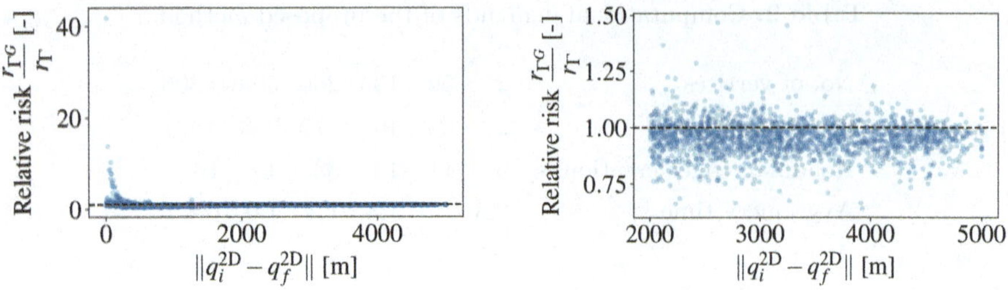

(a) All scenarios

(b) Scenarios where $\|q_i^{2D} - q_f^{2D}\| > 2000$ m

Fig. 5. The relative risk based on the distance between q_i and q_f: (**a**) all solved trajectories, (**b**) trajectories where q_i and q_f are further apart.

5% compared to the reference, and the 90% non-parametric interval of relative length is [0.73, 1.47]. The analysis of the relative risk based on the Euclidean distance between the start and final configurations is depicted in Fig. 5. Notice that the risk is increased mainly for the short scenarios, in which the maneuvers connecting q_i and q_f into the roadmap can significantly prolong the solution, and so the risk. On the contrary, the risk is reduced compared to the reference for the longer maneuvers.

5 Conclusion

A safe corridors roadmap generation for risk-aware trajectory planning is presented in the paper. The addressed problem is motivated by reducing the computational demands of the single-query risk-aware trajectory planning methods by determining a set of trajectories from which a safe corridors roadmap is created. The proposed method employs the GNG-based clustering to detect common low-risk trajectory segments, so-called safe corridors, among provided risk-aware trajectories. The resulting roadmap allows using graph-based planning for fast on-demand queries. The proposed method needs tens to hundreds of seconds to build the roadmap. However, the queries can then be solved within seconds, compared to hundreds of seconds required by the existing single-query risk-aware trajectory planners.

Acknowledgments. The presented work has been supported by the Ministry of Education Youth and Sports (MEYS) of the Czech Republic under project No. LTAIZ19013 The support of the Czech Science Foundation (GAČR) under research project No. 19-20238S is also acknowledged.

References

1. Bezanson, J., Edelman, A., Karpinski, S., Shah, V.B.: Julia: a fresh approach to numerical computing. SIAM Rev. **59**(1), 65–98 (2017)
2. Chitsaz, H., LaValle, S.M.: Time-optimal paths for a dubins airplane. In: IEEE Conference on Decision and Control (CDC), pp. 2379–2384 (2007)
3. Dijkstra, E.W., et al.: A note on two problems in connexion with graphs. Numer. Math. **1**(1), 269–271 (1959)
4. Facebook Connectivity Lab and Center for International Earth Science Information Network - CIESIN - Columbia University: High Resolution Settlement Layer (HRSL). Source imagery for HRSL © 2016 DigitalGlobe (2016). Accessed 10 Feb 2021
5. Fritzke, B.: A growing neural gas network learns topologies. Adv. Neural Inf. Process. Syst.**7** (1994)
6. Hasan, S.: Urban air mobility (UAM) market study. Technical report, National Aeronautics and Space Administration (NASA) (2019)
7. Melnyk, R., Schrage, D., Volovoi, V., Jimenez, H.: A third-party casualty risk model for unmanned aircraft system operations. Reliab. Eng. Syst. Saf. **124**, 105–116 (2014)
8. Moore, M.: 21st century personal air vehicle research. In: AIAA International Air and Space Symposium and Exposition: The Next 100 Years, p. 2646 (2003)
9. OpenStreetMap contributors: Planet dump retrieved from. https://planet.osm.org, https://ww.openstreetmap.org (2021). Accessed 10 Feb 2021
10. Primatesta, S., Rizzo, A., la Cour-Harbo, A.: Ground risk map for unmanned aircraft in urban environments. J. Intell. Robot. Syst. **97**(3), 489–509 (2019). https://doi.org/10.1007/s10846-019-01015-z
11. Sláma, J., Váňa, P., Faigl, J.: Risk-aware trajectory planning in urban environments with safe emergency landing guarantee. In: 2021 IEEE 17th International Conference on Automation Science and Engineering (CASE), pp. 1606–1612. IEEE (2021)
12. Váňa, P., Sláma, J., Faigl, J., Pačes, P.: Any-time trajectory planning for safe emergency landing. In: IEEE/RSJ International Conference on Intelligent Robots and Systems (IROS), pp. 5691–5696 (2018)

SOM Visualization Framework in Python, Including SOMStreamVis, a Time Series Visualization

Sergei Mnishko(✉) and Andreas Rauber

TU Wien, Vienna, Austria
sergei.mnishko@gmail.com, rauber@ifs.tuwien.ac.at

Abstract. The unique power of the Self-Organizing Map (SOM) lies specifically in the wide range of visualizations available to support explorative data analysis. However, these are usually scattered across independent libraries, making their integration and comparative application in data exploration difficult. Additionally, few visualizations focus on the dynamics of data forming a time series, while the importance of understanding data evolution is gaining importance in many settings. In this paper, we thus present SOMStreamVis, an extension to a visualization specifically tuned to capture and represent time-series information on self-organizing maps. We furthermore provide an PySOMVis, an open-source software framework realized in Python that integrates SOMStreamVis with a wide range of additional visualizations of the SOM, making it a powerful data exploration tool.

Keywords: Time series data · Visualization · Data exploration

1 Introduction

The Self-Organizing Map (SOM) is a powerful tool for data exploration. This power stems predominantly from the myriad of visualizations that allow interactive exploration of data, identifying clusters of different cardinalities or densities, understanding their hierarchical structure and semantic relationships as well as being able to characterize them by the attributes they share or that make them different. In order to put the SOM to good use, these visualizations need to be combined, and thus - ideally - available in integrated frameworks that allow us to put different visualizations side-by-side, to see in how far the information provided agrees or disagrees with specific hypotheses. However, while many visualizations have been proposed, few frameworks provide more than a fundamental set of these. We thus present, as the first key contribution of this paper, *PySOMVis*[1], a visualization framework for the SOM implemented in Python, that provides a range of visualization techniques (such as U-Matrix, Smoothed Data Histogram, Metro Map, Clustering approach, Chessboard Visualization, Neighborhood Graphs, SOMStreamVis and etc.), both as independent functions,

[1] https://github.com/smnishko/PySOMVis.

J. Faigl et al. (Eds.): WSOM+ 2022, LNNS 533, pp. 98–107, 2022.
https://doi.org/10.1007/978-3-031-15444-7_10

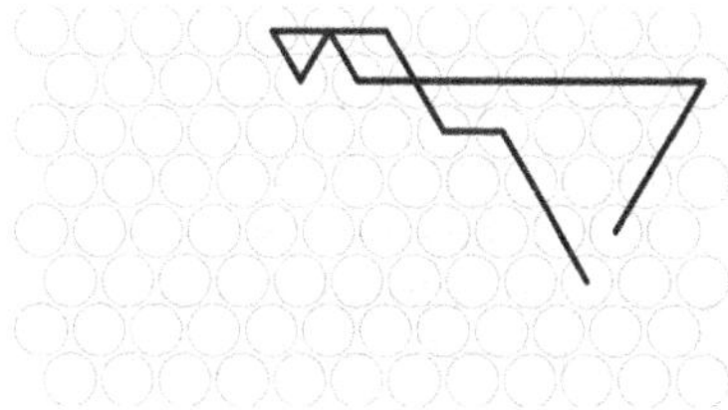

Fig. 1. BMU track in SOMPak package [4].

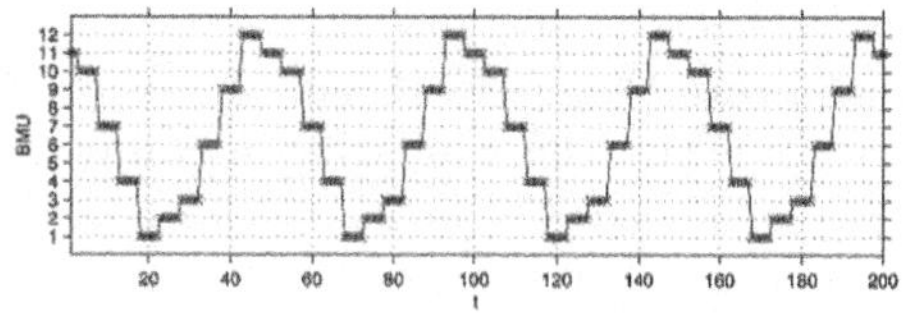

Fig. 2. T-axis visualization [6].

as interactive stand-alone application, or integrated into Jupyter notebooks. This allows flexible deployment and adoption in a wide range of interactive data exploration scenarios.

Analyzing data streams imposes additional restrictions on the algorithms and visualizations. One of the core tasks in understanding time series data is the identification of trends and fluctuation patterns over time. Three stream characteristics are of particular interest, namely (1) Concept drift; (2) Concept evolution; and (3) Pattern detection. This is hardly supported by any of the classical visualizations, frequently being limited to trajectory mappings. A few visualizations have been proposed that help with identifying such trends. We thus expand on existing visualizations tuned to time series data, adding *SOMStreamVis*, which visualizes trajectories of time series mapped onto SOMs, providing color coding to clearly distinguish cluster membership of various sections of the data.

The remainder of this paper is structured as follows: Sect. 2 provides a short overview of existing visualizations for the SOM as well as specific adaptations of the SOM to deal with time-series data. The principles of our time series visualization approach are presented in Sect. 3. This is followed by a brief presentation of the *PySOMVis* toolbox in Sect. 4 and an evaluation of our contributions in Sect. 5. Section 6 provides a summary of our contributions as well as an outlook on future work.

2 Related Work

2.1 Visualizations for the SOM

A large number of visualization techniques make the SOM into the powerful tool for data exploration that it is, revealing clusters, their distribution, and semantic characteristics. The visualizations can be divided into several categories, depending either on the data used (e.g., weight vectors of the trained map or also using training data information) or the type of information to be extracted (e.g., distance information between units, data density, topological arrangements) or the type of visualization applied (e.g., coloring the map, printing or drawing on the map). Some methods help to identify clusters, their similarities, borders, and attributes. For example, the *U-Matrix* [14] approach which interpolates the distance between neuron units, revealing mountains (boundaries) and valleys (clusters). Similar calculations are provided by the *D-Matrix* method, where the unit averaging the distance between its neighboring. Those methods are very popular and many Python visualizations provide them out of the box. Also, there are approaches that help to identify the similarity of components of neurons. The *Component Planes* approach slices the weight vectors, analyzing the correlation between components, by coloring the regions of the map with the high and low values. *MetroMaps* [9] summarizes compactly the information across a number of component planes. It uses the idea of a metro lines map, where lines represent the components and their location on the map representing similarities between the components.

Using input data can reveal density information, using, e.g., *Hit-histograms*, which colors the map according to a unit's popularity. *Smoothed Data Histogram* [10] is an extension smoothing over the n-best matching units (in a variety of weighting schemes) that reveals clusters at different levels of granularity and, depending on the type of smoothing applied, focuses on clusters of similar density or cardinality. The *P-matrix*[15] provides a visualization of density by showing the density within a certain radius around each weight vector in input space. The *U*-matrix* [16] visualizations combines the *P-matrix* and *U-matrix* approaches, and helps to identify additional structure inside the cluster of Emergent self organizing maps. Other techniques, such as *Sky-metaphor* [5], *Neighborhood Graphs* [11] also help in identifying the density and topology preservation of a trained map. Additionally, there are techniques, which reveal information about class distribution in supervised setting on the map, such as Pie Charts or Chessboard [8].

While there is a huge number of implementations for training SOMs in Python ([2,3,17,18] etc.), there is lack of implementations of the huge number of visualizations available. Most of these are only available as individual components in Matlab, R or Java, with very few frameworks providing a collection of these. One example of the latter is the Java SOMToolbox framework [7], which provides all of the visualizations listed above plus a few additional ones for comparing several SOMs. However, being implemented in Java it lacks in scalability and makes integration in modern Python-based data analytics pipelines difficult. We thus present *PySOMVis*, a Python-based framework, which tries to fill this gap.

2.2 Time Series Data and the SOM

The SOM has been used for exploration of data dynamics. One approach is *ubiSOM* [12]. It eliminates the time parameter for the training process (replacing it by the distance between unit and data instance) and identifies two additional metrics which control the force and quality of the training process. They are *average neuron utility*, which serves to identify missing classes and is based on the frequency of activation of each unit, and another parameter is *average quantization error*, which considers the exponential decreasing value during the training process. Whenever the metrics identify a missing class or abrupt change in income data, it changes the learning rate and neighborhood radius from fine-tuning mode to the initial values.

Apart from changes to the architecture and training process, there are visualization approaches that help to explore data dynamics. Trajectory-based visualization, such as depicted in Fig. 1, shows the sequence of best-matching units (BMU) for sequences of data instances on the map. These have been widely applied, e.g., for analyzing bankruptcy [1] or industrial process control [13]. Another visualization approach is suggested by Yonggang et al. [6], which we refer to as the time-axis (or T-axis) visualization. They applied the SOM to distinguish patterns of ocean currents. They depict the BMU sequence on a graph where the X-axis represents the sequence of input data, and the Y-axis represents the BMU's index (cf. Fig. 2). This visualization serves as the inspiration for *SOMStreamVis*, combining it with methods from static SOM visualizations to reveal additional insights on cluster membership.

3 SOMStreamVis

SOMStreamvis is based upon the principle of the T-axis visualization, i.e., it depicts the sequence of indices of the best-matching units over time. Different to the T-axis visualizations it does (1) apply smoothing over a time window of data instances to obtain a clearer representation of trends, eliminating minor BMU index fluctuations within a cluster; and (2) adds color-coded depictions of the actual BMU indices reflecting, e.g., cluster membership or distance to the neighboring units to reflect the closeness to a cluster boundary, depending on the base-visualization chosen.

Given a trained SOM with $k = x \times y$ units u, we create a linearized index i of units $1 \leq i \leq k$. This linearization is trivial in the case of a 1-dimensional SOM (i.e., either x or y being 1). In case of a rectangular SOM, the linearization happens via iterating first over the shorter dimension and doing so in a zig-zag mode across the longer dimension. (Note: for the visualization to produce meaningful results, we recommend using either a linear SOMs or SOMs with a high aspect ratio.) The *StreamVis* average line (sva) is then computed as:

$$sva(t) = \sum_{j=1}^{n-1} \frac{\beta}{n-1} \cdot i_{BMU}(t-j) + (1-\beta) \cdot i_{BMU}(t), \tag{1}$$

where

- $sva(t)$ - stream vis average line index position at time t

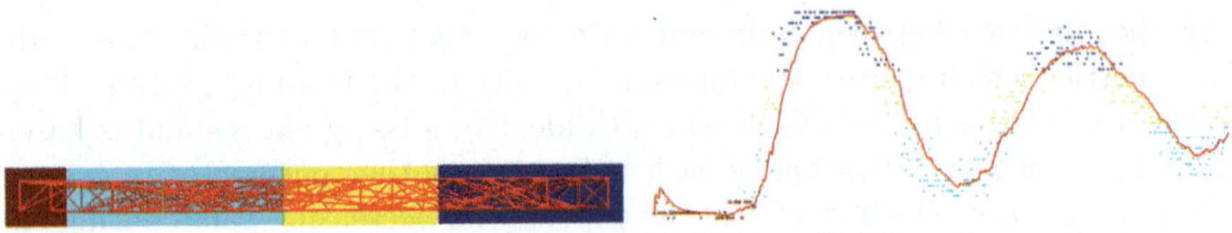

(a) BMU trajectory on SOM, structured into 4 clusters using WARD clustering (b) Colored BMUs and smoothed average in T-axis visualization

Fig. 3. 2×20 SOM of synthetic 3-d sinusoidal data colored by 4 clusters (WARD).

- $i_{BMU}(t)$ - index of the best-matching unit for the data instance presented at time t
- β - smoothing factor
- n - size of the sliding window, i.e., past BMU indices to consider.

It computes an exponentially weighted average between the indices of the n best matching units in a sliding window of n data instances. β controls the weight of the preceding BMUs' index values and thus leads to a smoother line plot for high values of β. Low values of β assign less weight to the preceding BMUs indices, following more closely the current BMU's index. Setting β to 1 results in the line of the T-axis visualization.

To reveal additional information *SOMStreamVis*, additionally plots the actual (i.e., non-averaged) index position of the BMU at each time step. This reveals the actual spread of BMU indices across the averaged line (which naturally is larger for larger SOMs, but also tends to produce larger jumps for SOMs with low aspect ratio as the linearization will usually break neighborhood relationships.) By coloring the BMU index positions in the plot we can reveal additional information such as cluster membership, or transition between clusters. To depict cluster membership, we can apply any color coding technique for the SOM that distinguishes clusters. Examples include a clustering of the SOM's weight vectors using, e.g., k-means or WARD clustering, or any hierarchical clustering method. Alternatively, the SOM can be colored using, e.g., the D-Matrix visualization, where each unit is colored according to the average distance of the unit's weight vector to the neighboring units' weight vectors. This reveals the closeness to a cluster boundary, and the magnitude of change as the trajectory of data mappings passes through various areas of the SOM.

To illustrate *SOMStreamVis*, we use the synthetically data of a 3-d sinusoidal signal with 2% of Gaussian noise. The data is mapped onto a 2×20 SOM with sigma $= \mathbf{1}$, learning rate $= \mathbf{0.8}$, over all **500** instances for **10,000** times, using the *miniSOM* library. Figure 3 shows both the SOM (a) and the *SOMStreamVis* approach (b). Here, WARD clustering is used to structure and color the SOM into 4 clusters. The classic trajectory visualization highlighted in red (Fig. 3(a)) over all input instances does not reveal any clear pattern or information of state in time. *SOMStreamVis* (Fig. 3(b)) reveals clear cyclic pattern via the BMU index positions. Moreover, the exponentially weighted average line emphasises the trend, while coloring the BMUs in the T-axis visualization helps to identify the state/SOM mapping.

4 PySOMVis Toolbox

There are many implementations of the Self-Organizing Map, both open-source and commercial, and in a variety of programming languages like C++, Java, Python, etc. Yet, there is a massive gap in visualization techniques. The most common (and frequently sole) visualization provided is the *U-matrix*. Some repositories contain *Starburst* [17], *Quantization error, Hit-histograms*, and *Component planes*. We thus first present *PySOMVis*, an open-source Python-based GUI implementation for analyzing trained SOMs. It provides a wide range of different visualizations for the SOM, such as *Chessboard Visualization, Clustering approach, Component Plane, D-Matrix, Hit Histogram, Metro Map, Neighborhood Graphs, Pie Chart, Smoothed Data Histogram, U-Matrix, U*-Matrix, P-Matrix, Quantization Error*, as well as the new *SOMStreamVis*.

The GUI provides compact visualization with the general controls of rotation, different coloring, interpolation, also it contains the visualization part and panel to control its parameters.

Figure 4 provides an overview of some of the visualizations provided. It provides different views of a **10** $\times$ **10** SOM trained with the Iris Dataset[2] from the UCI Machine Learning Repository, a 4-dimensional data set of describing 50 instances of each of three classes of iris flowers by their petal and sepal leaf length and width. These were compared to visualizations obtained from the Java-based SOMToolbox [7]. All methods present similar results apart from some slight discrepancy for the Metro-map visualization due to the heuristic that is used to calculate straight metro lines with 45 or 90 angles.

The implementation has been tested to scale to large SOMs (more than 1 million units). It provides an interactive interface to explore the data with different visualizations approaches, allowing to export plots in different formats. It can be conveniently integrated in Jupyter Notebooks, but also can be adapted as a stand-alone Dashboard.

5 Evaluation of Data Stream Visualization

We evaluate *SOMStreamVis* on climate data covering a timespan of 35 years, from 1985–2020. The data was obtained from the National Aeronautics and Space Administration (NASA) Langley Research Center (LaRC) Prediction of Worldwide Energy Resource (POWER) Project and was collected from the POWER Project's Hourly 2.0.0 version on 2021/11/21[3]. It consists of 5 features presented in Table 1.

[2] https://archive.ics.uci.edu/ml/datasets/iris.

[3] https://power.larc.nasa.gov/data-access-viewer/.

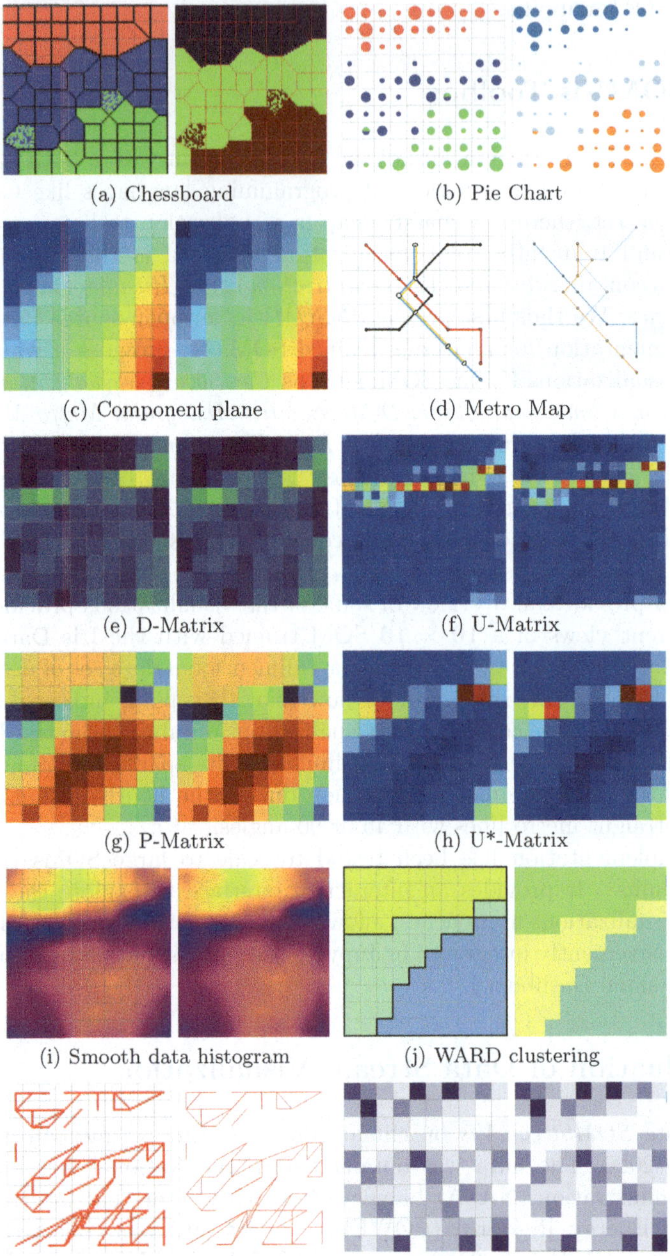
(a) Chessboard (b) Pie Chart
(c) Component plane (d) Metro Map
(e) D-Matrix (f) U-Matrix
(g) P-Matrix (h) U*-Matrix
(i) Smooth data histogram (j) WARD clustering
(k) K-Nearest Neighbor (l) Hit-histogram

Fig. 4. Different visualizations of PySOMVis. Comparison between Java (left) and Python (right) implementations.

(a) D-Matrix (b) Ward clustering (4 clusters)

Fig. 5. SOM coloring techniques on 2-years weather forecast.

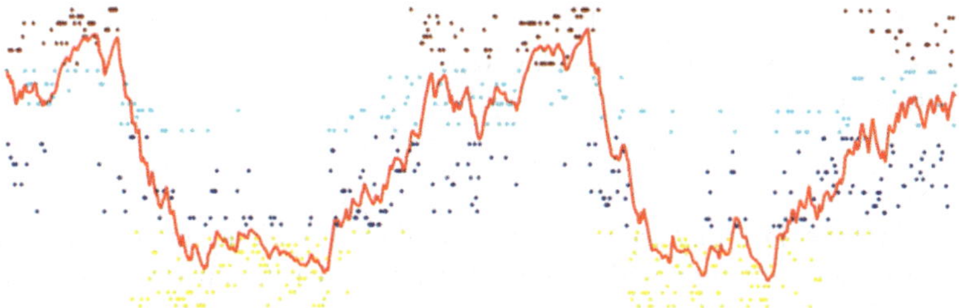

Fig. 6. SOMStreamVis visualization applied on 2-years weather forecast data.

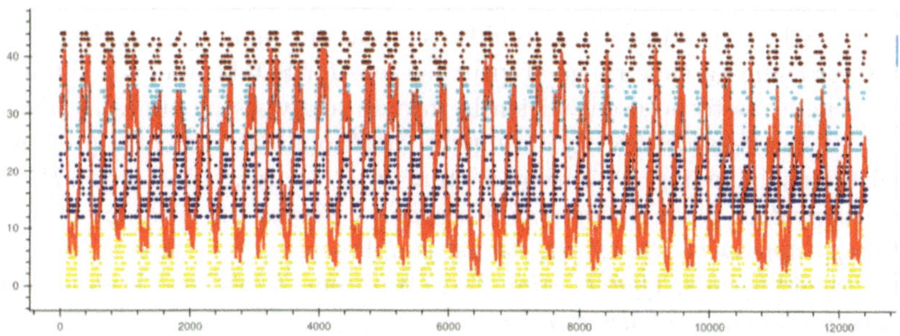

Fig. 7. SOMStreamVis visualization applied on 35-years weather forecast data.

Due to the seasonality, this data represents low and high temperature states (winter, summer) as well as interim states (autumn and spring). For training the SOM we first apply linear interpolation for missing values, then scale it by subtracting the median and scaling it to the quantile range using the *RobustScaler* of *scikitlearn*.

We train a **3** × **15** SOM using the Pyhton based *MiniSOM* library with the learning rate: **0.8**, neighborhood = **1.0**, iterations = **10,000**, randomize init = **8**.

The change between seasons is impossible to identify via standard static visualizations as depicted in Fig. 5, which shows the D-Matrix, and a clustering of the weight vectors using WARD clustering.

Applying *SOMStreamVis* we, first of all, obtain a clear depiction of the fluctuation and periodicity of the data on the SOM, both in the detailed 2-years analysis depicted in Fig. 6, as well as for the entire 35-year period depicted in Fig. 7. We see weather "extremes" not reoccurring every year as certain parts of the SOM (node indices in the range of 35 and above, colored in dark red), stemming from the rightmost column of the SOM in Fig. 5 do not as frequently

Table 1. NASA LaRC POWER Dataset

N	Feature name	Min	Max	Deviation
1.	Relative humidity at 2 m (%)	43.9	100	8.59
2.	Surface pressure (kPa)	94.88	103.43	1.01
3.	Temperature at 2 m (C)	−37.05	27.55	11.60
4.	Wind speed at 50 m (m/s)	0.64	17.64	2.32
5	Downward thermal infrared (longwave) Radiative flux ($kW - hr/[m^2]/day$)	3.14	9.84	1.22

serve as BMU for data instances between samples 1,500 and 3,000, where the red line of the smoothed timeline does not fluctuate to these top-sorted unit indices.

6 Conclusions and Future Work

We presented *PySOMVis*, a new SOM visualization framework implemented in Python, providing a wide range of visualizations for SOMs, allowing these powerful mechanisms to be used efficiently in data analytics pipelines as libraries, and interactive tool or directly in Jupyter notebooks. It currently provides 14 visualizations including *D-Matrix, U-Matrix, P-Matrix, U*-Matrix, Hit Histogram, Smoothed Data Histograms, Component Planes, Metro Maps, Neighborhood graph approach, Pie Chart, Chessboard Visualizations, several clusterings of unit weight vectors, Quantization Error and time series visualization such as Trajectory Plots and SOMStreamVis.*

We presented *SOMStreamVis*, an extension to the time axis visualization of BMU indices for depicting the sequence of states of a time series mapped onto a SOM, combining the trajectory of node indices with color coding to reveal cluster membership.

Currently, we are expanding the set of visualizations available in *PySOMVis*, adding support for metaphor graphics, supporting non-rectangular *Mnemonic SOMs*, tree and graph based cluster identifications such as *Minimum Spanning Trees* and *Cluster Connections*, as well as the visualization of quality measures capturing the topology preserving characteristics of a map. Furthermore, we are considering extensions to the time series representation by obtaining aligned views between *SOMStreamVis* and regular SOM visualizations, as well as investigating data streams on larger 2-dimensional maps as a 3-dimensional trajectory.

References

1. Chen, N., Ribeiro, B., Vieira, A., Chen, A.: Clustering and visualization of bankruptcy trajectory using self-organizing map. Expert Syst. Appl. **40**(1), 385–393 (2013). https://doi.org/10.1016/j.eswa.2012.07.047

2. Hanke, M., Halchenko, Y.O., Sederberg, P.B., Hanson, S.J., Haxby, J.V., Pollmann, S.: PyMVPA: a python toolbox for multivariate pattern analysis of FMRI data. Neuroinformatics **7**, 37–53 (2009). https://doi.org/10.1007/s12021-008-9041-y
3. Juha, V., Johan, H., Esa, A., Juha, P.: SOM toolbox for Matlab 5. Technical report, Helsinki University of Technology (2000). http://www.cis.hut.fi/somtoolbox/package/papers/techrep.pdf. Accessed 10 May 2022
4. Kohonen, T., Hynninen, J., Kangas, J., Laaksonen, J.: Som_pak: the self-organizing map program package. Technical report A31, Helsinki University of Technology, Finland (1996)
5. Latif, K., Mayer, R.: Sky-metaphor visualisation for self-organising maps. In: International Conference on Knowledge Management (I-KNOW 2007), pp. 400–407. Graz, Austria (2007)
6. Liu, Y., Weisberg, R.H., Mooers, C.N.K.: Performance evaluation of the self-organizing map for feature extraction. J. Geophys. Res. Oceans **111**(C5) (2006). https://doi.org/10.1029/2005JC003117
7. Mayer, R., Rauber, A.: The Java SOMToolbox @ IFS, TU Vienna. http://www.ifs.tuwien.ac.at/dm/somtoolbox/index.html. Accessed 10 May 2022
8. Mayer, R., Aziz, T.A., Rauber, A.: Visualising class distribution on self-organising maps. In: de Sá, J.M., Alexandre, L.A., Duch, W., Mandic, D. (eds) ICANN 2007. LNCS, vol. 4669. Springer, Heidelberg (2007). https://doi.org/10.1007/978-3-540-74695-9_37
9. Neumayer, R., Mayer, R., Polzlbauer, G., Rauber, A.: The metro visualisation of component planes for self-organising maps. In: IEEE International Joint Conference on Neural Networks (IJCNN 2007), pp. 2788 – 2793 (2007). https://doi.org/10.1109/IJCNN.2007.4371401
10. Pampalk, E., Rauber, A., Merkl, D.: Using smoothed data histograms for cluster visualization in self-organizing maps. In: Dorronsoro, J.R. (ed.) Artificial Neural Networks – ICANN 2002, pp. 871–876. Springer, Berlin Heidelberg (2002)
11. Pölzlbauer, G., Rauber, A., Dittenbach, M.: Advanced visualization techniques for self-organizing maps with graph-based methods. In: Wang, J., Liao, X.-F., Yi, Z. (eds.) ISNN 2005. LNCS, vol. 3497, pp. 75–80. Springer, Heidelberg (2005). https://doi.org/10.1007/11427445_13
12. Silva, B., Marques, N.C.: The ubiquitous self-organizing map for non-stationary data streams. J. Big Data **2**(1), 27 (2015). https://doi.org/10.1186/s40537-015-0033-0
13. Simula, O., Alhoniemi, E., Hollmen, J., Vesanto, J.: Analysis of complex systems using the self-organizing map. In: Proceedings of 4th International Conference on Neural Information Processing (ICONIP'97), pp. 1313–1317. Dunedin, New Zealand (1997)
14. Ultsch, A.: Self-organizing neural networks for visualisation and classification. In: Opitz, O., Lausen, B., Klar, R. (eds.) Inf. Classif., pp. 307–313. Springer, Berlin Heidelberg (1993). https://doi.org/10.1007/978-3-642-50974-2_31
15. Ultsch, A.: Maps for the visualization of high-dimensional data spaces. In: In Proceedings Workshop on Self-Organizing Maps (WSOM 2003), pp. 225–230. Kyushu, Japan (2003)
16. Ultsch, A.: U*Matrix 1 U*-Matrix: a tool to visualize clusters in high dimensional data (2004)
17. Vettigli, G.: MiniSom (2021). https://github.com/JustGlowing/minisom. Accessed 10 May 2022
18. Wittek, P., Gao, S.C., Lim, I.S., Zhao, L.: Somoclu: an efficient parallel library for self-organizing maps. J. Stat. Softw. **78**, 1–21 (2017). https://doi.org/10.18637/jss.v078.i09

Visual Insights from the Latent Space of Generative Models for Molecular Design

Salvatore Cavallaro[1,2], Alfredo Vellido[1,3], and Caroline König[1](✉)

[1] Computer Science Department, and Intelligent Data Science and Artificial Intelligence (IDEAI -UPC) Research Center, Universitat Politècnica de Catalunya, Barcelona, Spain
{avellido,ckonig}@cs.upc.edu
[2] Università degli Studi di Catania, Catania, Italy
salvatore.cavallaro7@studium.unict.it
[3] Centro de Investigación Biomédica en Red, Bioingeniería, Biomateriales y Nanomedicina (CIBER-BBN), Cerdanyola del Vallès, Spain

Abstract. The inverse design of materials is an active field of research in engineering, for which generative models are eliciting an increasing interest, due to their ability to create new possibly meaningful representations of materials from the explored space of examples. In this brief study, we investigate the relative usefulness of different representations of small organic molecules using a variational autoencoder model. Exploratory visualization of the latent space of the model is used to increase the interpretability of the results. Visualization is also used to assist the assessment of the reconstruction quality of the model.

Keywords: Variational autoencoders · latent space · inverse design of organic molecules

1 Introduction

The engineering of new materials is a relevant technological advance in current industrial processes. The artificial and inverse design of materials, in particular, holds the promise of accelerating the exploration and discovery of materials with improved properties. Artificial Intelligence and Machine Learning methods enable a smart exploration of a possibly very large compound space, making the discovery of interesting compounds with improved features feasible in a shorter time [1,2].

Riding the wave of advances in deep learning, generative models have risen as an approach for inverse design due to their ability to generate new representations from a learnt probability distribution of the data space [3]. Variational Autoencoders (VAEs) [4] and Generative Adversarial Networks (GANs) [5] are two types of generative networks used in different types of applications, such

C. König—This work is funded by Spanish PID2019-104551RB-I00 research project.

J. Faigl et al. (Eds.): WSOM+ 2022, LNNS 533, pp. 108–117, 2022.
https://doi.org/10.1007/978-3-031-15444-7_11

as for the generation of artificial text [6,7], image creation [8], or, closer to the current work, molecular design [9,10], to name a few.

VAEs use an *encoder* and *decoder* network for encoding and decoding the original data in and from a *latent space* of lower dimension, thus learning a representative probability distribution of the data in this space through dimensionality reduction and abstraction. GANs perform adversarial training for model creation, also using two networks, namely a *generator* and a *discriminator*. The former is trained to learn the sampling of instances from the latent space, while the latter verifies the authenticity of generated items and aims at improving the capabilities of the network in the generation of the characteristics.

In this brief study, we investigate the ability of VAEs to learn representations of small organic molecules. We specifically aim to assess, through VAE, what is the best molecular representation for the generative task. The model is trained using variants of the QM7 dataset that include the description of the quantum-mechanical properties of small organic molecules from their chemical space [11]. An exploratory analysis of the created VAE model for the assessment of the quality of reconstruction of the model and its generative capabilities is carried out. Visualization techniques are used as model explainability-support tools to provide users with insights about important aspects of the molecular representations, including the reconstruction quality, as well as the data distribution of the data in the latent space. For the latter, clustering of the molecules in the original space is carried out to verify whether the similarity between compounds in the original dataspace is preserved in the latent space of the VA.

2 Materials

The specific datasets chosen to illustrate our approach are **QM7** and **QM7b**, which are subsets of the GDB-13 [12] with about 1 billion organic molecules.

QM7 is composed of all the molecules consisting of up to 23 atoms, including 7 heavy atoms, C, N, O and S, and totalling 7165 molecules. The dataset provides different features, the most relevant of which are the Coulomb matrix representation, the atomization energies of each molecule, the atomic charges and the cartesian coordinates of each of their constituting atoms. More specifically, the Coulomb matrix [13] is a global descriptor representing the electrostatic interaction between atoms; it is a symmetric matrix, with elements $C_{ii} = \frac{1}{2}Z_i^{2.4}$ and $C_{ij} = \frac{Z_i Z_j}{|R_i - R_j|}$, where Z_i is the nuclear charge and R_i is the position of each atom. This structure is invariant to rotation or translation of the molecule and the elements on its diagonal represent the interaction of an atom with itself.

The QM7 dataset includes different entries: X (of dimensions $7165 \times 23 \times 23$) as the Coulomb matrix, T (7165×1) as the atomization energy, Z (7165×23) as the atomic charges and R ($7165 \times 23 \times 3$) as the cartesian coordinates.

QM7b is another subset of GDB-13 and is further described in [14]. It contains a description based on a Coulomb matrix and 14 additional properties, namely: atomization energy E, static polarizabilities α, frontier orbital eigenvalues HOMO and LUMO, ionization potential IP, electron affinity EA, first excitation energy $E^*_{1^{st}}$, excitation of maximal optimal absorption E^*_{max} and the corresponding intensity I_{max}. These properties are predicted at three levels: ZINDO (Z), PBE0 (P) and GW (G). It is an extension of QM7 dataset and contains 7211 molecules.

The goal of this work is the creation of VAE models from the information in the QM7 and QM7b datasets. We aim to ascertain which of these molecular descriptions is the most adequate to be used for the generative approach to inverse design of molecules. A series of data pre-processing steps are required. For QM7, all data structures have to be merged into a unique set. The result is a concatenation of X, Z, R and T arrays yielding a dataset of dimensions $7165 \times 23 \times 28$, where 7165 is the number of molecules, 23 is the number of atoms per molecule and 28 is the number of features per atom. A similar approach is followed with QM7b, where X and T arrays from the dictionary are merged. Its dimensions are $7211 \times 23 \times 37$. Finally, data are normalized to bring all values into the range $[0, 1]$.

3 Methods

3.1 Variational Autoencoder

Model Architecture: The VAE consists of two blocks: an encoder, used to compress the input and produce the representation into the latent space z, and a decoder, used to decompress the representation from the latent space z and produce the content which reconstructs the original sample. Figure 1a depicts the network architectures for the VAE. A particular aspect of the VAEs used as generators is that the encoder is used only for training purposes, then it is discarded.

Encoder: The encoder is composed by a sequence of layers, each of which processes the input coming from the previous one and provide the output to feed the next one. The first is an *input layer*, which is a tensor (Generalization in n-dimensional space of vector/matrix) whose shape is parametrized according to the type of the dataset used to perform the training phase. The QM7 encoder input layer is $23 \times 28 \times 1$ while QM7b's is $23 \times 37 \times 1$. The second, third and fourth[1] layers are *convolutional layers* characterized by the parameters in Fig. 1b. Then, the output of the convolutional layers is flattened to make it usable by the 1-dimensional fully connected section of the neural network. The next layer is a dense one with 16 neurons and a ReLU activation function. Then, there are two more dense layers, containing a number of neurons equal to the latent space size, which in this case is 2 for visualization purposes. They are used to learn mean (μ) and covariance (Σ) matrices of the distribution representing input samples into the latent space. The output of these two layers is finally processed by yet another layer that performs random sampling on the distribution defined by these values of μ and Σ to finally obtain the latent representation of the input sample.

[1] Only for QM7b.

Decoder: The decoder has a similar structure to the encoder, but performing opposite operations. The first one is a tensor input layer which takes the 2-D representation from the latent space. The second layer is again a dense one and its output is reshaped to be suitable for processing by the next 2-dimensional layers, which are *transposed convolutional layers*; they are also described by the parameters in Fig. 1b. The last layer uses a sigmoid activation function instead of a ReLU, to obtain the reconstructed sample from the latent representation in z.

Model Hyperparameters: Several hyperparameters must be taken into account in model training. First, an adequate **loss function** must be chosen. It includes a *reconstruction loss* term and the *Kullback-Leibler (KL) divergence* term:

$$loss = ||x - \hat{x}||^2 + KL[N(\mu_x, \sigma_x), N(0, I)]$$

The *reconstruction loss* is defined as the Mean Square Error (MSE) between the original data x and the reconstructed sample $\hat{x}$, as the data is continuous. The total MSE from the decoder is computed according to $\frac{1}{n}\sum_{i=1}^{n}(x_i - \hat{x}_i)^2$. The

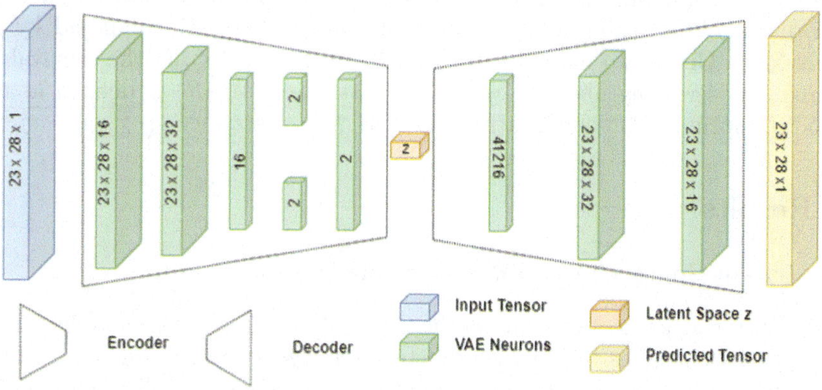

(a) Architecture QM7 dataset

Hyper-parameter	Value
Number of filters (QM7)	16/32 (2 Layers)
Number of filters (QM7b)	16/32/64 (3 Layers)
Filter Shape	3 × 3
Stride	1
Padding	same
Activation Function	ReLU

(b) Hyper-parameters of Convolutional Layers

Fig. 1. VAE architectures for the QM7 and QM7b dataset

second term, the *KL divergence*, is used to regularize the aproximation of the data distribution, as it measures the difference between the approximation $q_x(z)$ of the latent variable z and the exact probability distribution of $p(z|x)$. Therefore, the KL term is responsible for mapping similar datapoints into contiguous regions in the latent space.

The Adam optimizer [15] was used for training the models. A related hyperparameter is the *learning rate*, a value that determines how big is the step towards convergence of the loss function to a minimum. Its value is set to 0.001. A *batch size* of 128 was defined, so that the training takes into consideration batches of 128 samples from the original dataset and for each of them computes the mean gradient to update weights of the network. A related hyperparameter is the *number of epochs*, which was set to 70.

3.2 Clustering of Molecules

In this study, we cluster the molecular data using the *HDBScan* algorithm [16], as we have no prior knowledge about the expected *number of clusters.*

The objective is to verify whether the similarity between molecules using its original representation is kept in its representation in the latent space. According to [13], the distance between two molecules can be calculated with a metric based on its Coulomb matrix representation. This similarity metric is defined on a diagonalized version of the Coloumb matrix (on a set composed by its eigenvalues). Given two matrices M and M', where ϵ and ϵ' are the eigenvalues ordered by decreasing absolute value, the distance $d(M, M')$ is defined as the Euclidean norm of Coulomb matrices as $d(M, M') = d(\epsilon, \epsilon') = \sqrt{\sum_l |\epsilon_l - \epsilon'_l|^2}$.

4 Results

4.1 Latent Space Clustering and Visualization

A first clustering process was run on the Coulomb matrices of the complete QM7 dataset. HDBScan was tested with different settings for its parameters and the chosen ones were *min_cluster_size* = 5 and *min_samples* = 5. Different similarity metrics were tested to get a variety of clustering results and to compare them.

We aim to verify whether the similarity of the molecules is preserved in its representation of the latent space, i.e., whether similar molecules are mapped in contiguous regions in the latent space. Figure 2a shows an example of the visualization of the data in VAE latent space, where clusters obtained with the Jaccard (M_3) metric from the HDBSCAN algorithm are depicted with different colours. In this example, we see how the clustering found a well-defined grouping. Interestingly, it exactly reflects the number of atoms in the molecules. Furthermore, it is similar to the results obtained with 6 other different metrics, namely Canberra (M_1), Dice (M_2), Jaccard (M_3), Kulsinski (M_4), Matching (M_5), Rogerstanimoto (M_6) and Sokalsneath (M_7). We therefore can confirm that similarity in the original space is preserved in the latent space. A further way of evaluating the metrics is through the comparison of their performances. In Fig. 2b, we compare the number of clusters and the number of data classified as outliers by HDBSCAN, which are data points that the algorithm refuses to assign to a cluster.

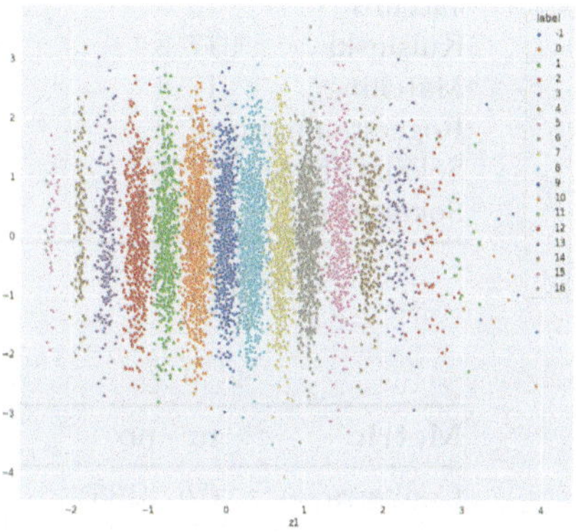

(a)

Metric	nc	no
Canberra	18	695
Dice	17	8
Jaccard	17	8
Kulsinski	18	3
Matching	17	8
Rogerstanimoto	17	8
Sokalsneath	17	8

(b)

Fig. 2. (a) Clustering result for *QM7* visualized in VAE latent space; (b) table including metric, clusters (nc), outliers (no).

Clustering of QM7b followed the same approach and same settings as for QM7. The visualization of clusters reveals worse performance as the groups are not so clearly separated. Most metrics did not yield well defined and homogeneous clusters for the compounds mapped into the latent space z. Figure 3b shows the visualization of clusters obtained by the Kulsinski metric. These results are comparable to that of other 7 metrics. Figure 3b shows a summary of all these information. The lower performances of clustering in this second case are probably related to the need of using a set of different molecular properties for the QM7b dataset, which justifies the experiments reported next using quantum-mechanical features.

The HDBSCAN clustering based on 14 quantum-mechanical properties of QM7b reported in Fig. 3d reveals that no straightforward data grouping was found. For most metrics, the algorithm found only two very large clusters, with the exception of the Euclidean and Manhattan metrics, which yielded 8 clusters.

We therefore decided to expand the experiment to merge the eigenvalues of the Coulomb matrix with the 14 quantum-mechanical properties for each molecule. Figure 3f describes the clustering results for several metrics. Interestingly, the results are quite heterogeneous. The Euclidean, Canberry and Manhattan metrics yield a high number of outliers and only a very small number of clusters. All the other metrics yield 15 clusters with a very low number of outliers. Figure 3e shows the distribution of the 15 clusters produced using the Kulsinski metric.

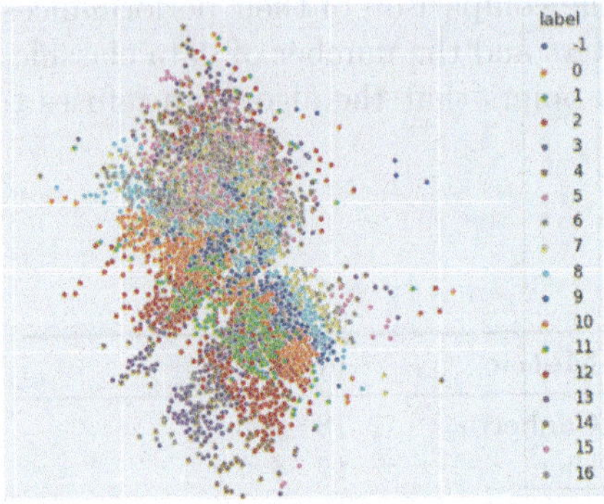

(a) Kulsinski metric

Metric	nc	no
Canberra	19	1343
Dice	18	3
Jaccard	18	3
Kulsinski	17	8
Matching	17	8
Rogerstanimoto	17	8
Sokalmichener	17	8
Sokalsneath	18	3

(b)

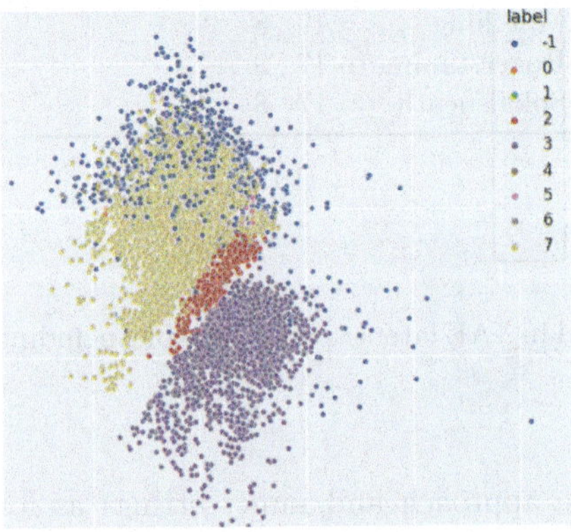

(c) Euclidean metric

Metric	nc	no
Canberra	62	3135
Dice	2	509
Euclidean	8	1034
Jaccard	2	509
Kulsinski	2	509
Manhattan	8	1122
Rogerstanimoto	2	509
Sokalmichener	2	509
Sokalsneath	2	509

(d)

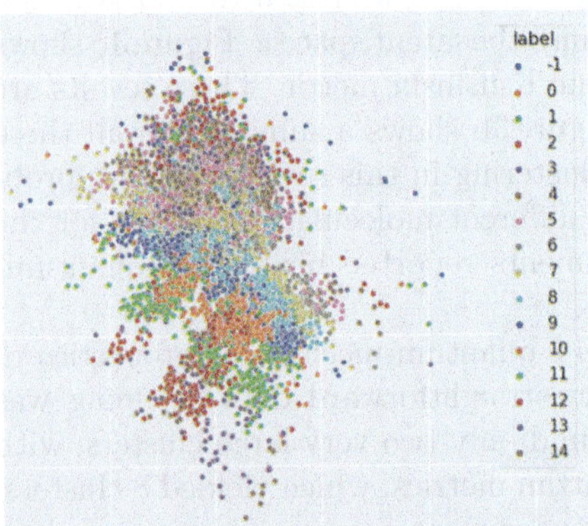

(e) Kulsinki metric

Metric	nc	no
Canberra	2	1506
Dice	2	20
Euclidean	2	564
Jaccard	15	20
Kulsinski	15	20
Manhattan	4	1752
Rogerstanimoto	15	26
Sokalmichener	15	26
Sokalsneath	15	20

(f)

Fig. 3. Clustering results for QM7b for different representations of the molecules: (a) and (b) Coulomb matrix eigenvalues; (c) and (d) 14 quantum-mechanical properties; (e) and (f) all features of QM7b. Displays on the left hand side show clusters according to a given metric. Tables at the right hand side show results describing the number of clusters (nc) and outliers (no) for each metric.

4.2 Reconstruction Quality Evaluation

Another objective of the study is to analyze the quality of reconstruction of the model. It measures how accurate the VAE model is at taking a real example from the original dataset, encoding it into the latent space z and reconstructing it through the decoder to get a sample as similar as possible to the original one. This assessment is necessary to ensure that the trained model works properly as a *generative model* able to randomly sample data from the latent space reasonably decoded as a new real molecule. This is relevant during the construction of the model and when generating a new instance from the model.

In the first case, we compute a matrix of MSE for each of the different 23×28 QM7 features and 23×37 QM7b features. The resulting matrix is then plotted as a heatmap. The picture shows darker/lighter squares for features with lower/higher reconstruction errors; therefore, it is a visual representation of the accuracy of the model at reconstructing each single feature.

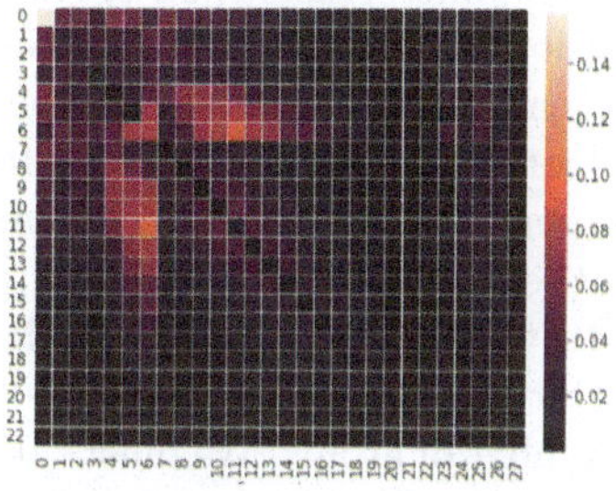

(a) Validation loss at the QM7 dataset

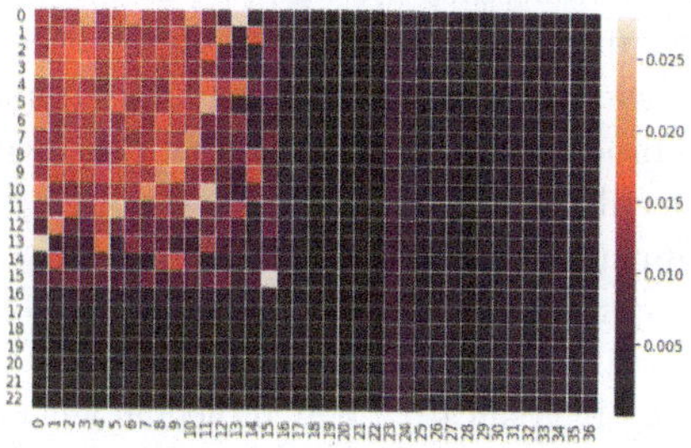

(b) Validation loss at the QM7b dataset

Fig. 4. Validation loss of the models as heatmap. The x-axis depicts the number of features and y-axis the number of atoms.

Figure 4a shows the result of this evaluation for QM7, where the square matrix identified by all the rows and columns 0–22 is the Coulomb matrix. These features are on average reconstructed better on the diagonal line[2], while the error is mostly concentrated around interaction between atoms 4–6 and 8–13. The columns 23–27 contain information about reconstruction of atomic charges Z, Cartesian coordinates R and atomization energy T. It is possible to observe that all the properties seem to be adequately reconstructed.

The same evaluation is performed for the model trained on the QM7b dataset. The heatmap in Fig. 4b shows a larger matrix because of the higher number of properties contained in the dataset. The features in the square sub-matrix ranging 0–22 are still the representation of the Coulomb matrix. In general, we see a lower reconstruction error than in the QM7 heatmap; nevertheless the highest amount of error is on the features 0–15. This behaviour could be a proof

[2] Interaction of the atom with itself, see Sect. 2.

of a higher concentration of molecules with a number of atoms less than or equal to 16. The remaining columns of the dataset (23–36) contain the properties described in Sect. 2.

Note that a similar heatmap-based visualization could be used to assess the reconstruction of an *individual molecule* from either QM7 or QM7b. We do not show any such representation in this study for the sake of brevity.

5 Discussion and Conclusions

The main objective of the study was the use of visualization to gain insights about the workings of the generative modeling in the process the generation of new instances (representations of small organic molecules) from the latent space. Further attention is paid to the quality of reconstruction. For the first objective, we used a combination of data clustering in the original data space with the visualization of the groupings in the VAE latent space, so as to verify whether similar molecules are mapped into contiguous regions. The preservation of similarity in the latent space is important as the generation of new instances relies on the nearby sampling from the latent space for a given instance. This means that the model will generate a new instance from the contiguous regions where the molecule was mapped by the encoder. From the experiments in Sect. 4.1, we can confirm that, for QM7, the latent space preserved the similarity of molecules, as the spatial distribution of the data corresponds to the groupings of molecules according to their *number of atoms.* This means that the generative model will probably generate a new molecule with the same number of atoms. We must take into account that QM7 contains information related to the electronic interactions of the atom, but no derived physico-chemical properties. This is important, as the other dataset, QM7b, contains more diverse information. The experiments on its latent space visualization yielded quite different results: the data clustered in the observed space were not separated into clearly differentiated groups in the latent space. An exploration of the groupings revealed that the representation of the quantum-mechanical properties seems to be relevant for the similarity in latent space, as the grouping according to the number of atoms (eigenvalue representation of the Coloumb matrix) is not preserved in the latent space, but the groupings of the quantum-mechanical properties yielded four contiguous regions. This means that this second model will probably generate a new molecule with similar properties to the sample, but *not only* taking into account the number of atoms.

We can conclude that the visualization of the VAE latent space allows us to assess which features of the dataset are best preserved in the data distribution of the latent space. It has been shown that a more diverse feature set yields a more complex representation in latent space, and that a deeper analysis of the features makes it possible to explain their relevance from the latent space representation. The experiments in Sect. 4.2 revealed for which features the model encountered difficulties. This information could be useful for the decisions on the choice of features of the molecular representation for inverse design processes, as a wide range of molecular representations are available, but not all of them are necessarily useful.

References

1. Schleder, G.R., Padilha, A.C., Acosta, C.M., Costa, M., Fazzio, A.: From DFT to machine learning: recent approaches to materials science-a review. J. Phys.: Mater. **2**(3), 032001 (2019)
2. Yang, D., Lv, J., Zhao, X., et al.: Functionality-directed first-principles screening of hybrid organic-inorganic perovskites with target intrinsic photovoltaic functionalities. Chem. Mater. **29**, 524 (2017)
3. Sanchez-Lengeling, B., Aspuru-Guzik, A.: Inverse molecular design using machine learning: generative models for matter engineering. Science **361**(6400), 360–365 (2018)
4. Kingma, D.P., Welling, M.: Auto-encoding variational bayes. In: 2nd International Conference on Learning Representations, ICLR 2014, Banff, 14-16 April 2014 (2014)
5. Goodfellow, I., Pouget-Abadie, J., Mirza, M., et al.: Generative adversarial nets. In: Advances in Neural Information Processing Systems, vol. 27 (2014)
6. Kim, J., Kong, J., Son, J.: Conditional variational autoencoder with adversarial learning for end-to-end text-to-speech. In: International Conference on Machine Learning, pp. 5530–5540 (2021)
7. Zhang, Y., Gan, Z., Carin, L.: Generating text via adversarial training. In: NIPS Workshop on Adversarial Training, Vol. 21, pp. 21–32 (2016)
8. Creswell, A., White, T., Dumoulin, V., et al.: Generative adversarial networks: an overview. IEEE Signal Process. Mag. **35**(1), 53–65 (2018)
9. Nesterov, V., Wieser, M., Roth, V.: 3DMolNet: a generative network for molecular structures. arXiv preprint. arXiv:2010.06477 (2020)
10. Maziarka, Ł, Pocha, A., Kaczmarczyk, J., Rataj, K., Danel, T., Warchoł, M.: Mol-CycleGAN: a generative model for molecular optimization. J. Cheminformatics **12**(1), 1–18 (2020). https://doi.org/10.1186/s13321-019-0404-1
11. Hoja, J., Medrano Sandonas, L., Ernst, B.G., et al.: QM7-X, a comprehensive dataset of quantum-mechanical properties spanning the chemical space of small organic molecules. Sci. Data **8**(1), 1–11 (2021)
12. Blum, L.C., Reymond, J.L.: 970 million druglike small molecules for virtual screening in the chemical universe database GDB-13. J. Am. Chem. Soc. **131**(25), 8732–8733 (2009)
13. Rupp, M., Tkatchenko, A., Müller, K.R., Von Lilienfeld, O.A.: Fast and accurate modeling of molecular atomization energies with machine learning. Phys. Rev. Lett. **108**(5), 058301 (2012)
14. Montavon, G., Hansen, K., Fazli, S., et al.: Learning invariant representations of molecules for atomization energy prediction. In: Advances in Neural Information Processing Systems, vol. 25 (2012)
15. Kingma, D.P., Ba, J.: Adam: A method for stochastic optimization. arXiv preprint. arXiv:1412.6980 (2014)
16. Campello, R.J.G.B., Moulavi, D., Sander, J.: Density-based clustering based on hierarchical density estimates. In: Pei, J., Tseng, V.S., Cao, L., Motoda, H., Xu, G. (eds.) PAKDD 2013. LNCS (LNAI), vol. 7819, pp. 160–172. Springer, Heidelberg (2013). https://doi.org/10.1007/978-3-642-37456-2_14

Author Index

J. Faigl et al. (Eds.): WSOM+ 2022, LNNS 533, p. 119, 2022.
https://doi.org/10.1007/978-3-031-15444-7

The manufacturer's authorised representative in the EU is Springer Nature Customer Service Centre GmbH, Europaplatz 3, 69115 Heidelberg, Germany. If you have any concerns regarding our products, please contact ProductSafety@springernature.com

Printed and bound by CPI Group (UK) Ltd, Croydon, CR0 4YY
30/09/2026
02225261-0009